MILWAUKEE IN STONE AND CLAY

MILWAUKEE IN STONE AND CLAY

A Guide to the Cream City's
Architectural Geology

Raymond Wiggers

NORTHERN ILLINOIS UNIVERSITY PRESS
AN IMPRINT OF CORNELL UNIVERSITY PRESS   ITHACA AND LONDON

First published 2024 by Cornell University Press
Printed in the United States of America

Library of Congress Cataloging-in-Publication Data

Names: Wiggers, Raymond, 1952– author.
Title: Milwaukee in stone and clay : a guide to the cream city's architectural geology / Raymond Wiggers.
Description: Ithaca [New York] : Northern Illinois University Press, an imprint of Cornell University Press, 2024. | Includes bibliographical references and index.
Identifiers: LCCN 2023024097 (print) | LCCN 2023024098 (ebook) | ISBN 9781501774645 (paperback) | ISBN 9781501774652 (ebook) | ISBN 9781501774669 (pdf)
Subjects: LCSH: Building materials—Wisconsin—Milwaukee—Guidebooks. | Stone buildings—Wisconsin—Milwaukee—Guidebooks. | Building stones—Wisconsin—Milwaukee—Guidebooks. | Architecture—Wisconsin—Milwaukee—Guidebooks. | Geology—Wisconsin—Milwaukee—Guidebooks. | Milwaukee (Wis.)—Buildings, structures, etc.—Guidebooks.
Classification: LCC TA425 .W56 2024 (print) | LCC TA425 (ebook) | DDC 691/.20977595—dc23/eng/20230608
LC record available at https://lccn.loc.gov/2023024097
LC ebook record available at https://lccn.loc.gov/2023024098

For my college students of two decades

All architecture is what you do to it when you look upon it,
(Did you think it was in the white or gray stone? or the lines of the
arches and cornices?)

—Walt Whitman, from "A Song of Occupations," in *Leaves of Grass*

Contents

Maps and Figures

Maps

Figures

Acknowledgments

This book, like its predecessor in the *Stone and Clay* series, is by no means simply mine. In fact it's been a collaborative effort involving, besides the lucky fellow who gets to have his name on the cover, a wonderfully talented team of Northern Illinois University Press and Cornell University Press staff members and contracted freelancers. Always at the top of my gratitude list is my primary contact, Senior Acquisitions Editor Amy Farranto, who skillfully shepherded the project through the review, manuscript-submission, and cover-design process—and who aided and abetted my hope that my first, Chicago volume would soon be joined by this Cream City sibling.

For the book's production phase, I was especially fortunate to once again work with Assistant Managing Editor Karen Laun, with whom I'd already forged this series format. Her superb layout sense was matched by her language skills and patience in answering all my questions about usage and style. In addition, a tip of my cap is also due to copy editor Glenn Novak, and to top-notch cartographer Daniel Huffman, whose proficiency and efficiency made the complex task of creating this volume's maps and geologic time scale look easy indeed.

The researching and writing of this book also confirmed my long-standing impression that Milwaukee and its county have more than the normal share of persons who possess that rare combination of kindness, openhearted helpfulness, and solid expertise. This secret should be carefully guarded lest the city be flooded by agents from other Midwestern communities seeking to even the odds. Blessed am I that I chose Milwaukee for my subject.

Because this work seeks to expose the remarkable connections between the science of geology and the pragmatic art of architecture, it was crucial that I found experts in each of these disciplines willing to help me with both key concepts and the nagging details. In the first-named discipline, I have been especially fortunate to benefit from the urban-geology savvy, insight, and keen moral support for this series provided by Renee Wawczak, professional geologist with the US Environmental Protection Agency. Kenneth C. Gass, noted Milwaukee-area paleontologist and honorary curator at the Milwaukee Public Museum Department of Geology, kindly reviewed my sections on this region's Silurian reefs and Estabrook Park's Devonian geology. From Esther Stewart, of the Wisconsin

Geological and Natural History Survey, I gleaned some interesting new facts about Badger State geology, and I was also assisted by the following Precambrian and Phanerozoic specialists at the Minnesota Geological Survey: Amy Rada-kovich Block, Julia Steenberg, Terry Boerboom, and Andrew Retzler. Separately, thanks to Todd Thompson, director of the Indiana Geological and Water Survey, and to his colleague Jennifer Lanman, archivist and collection manager, who was instrumental in helping me obtain some of this book's artwork. I also appreciate the efforts of Peter Lemiszki, chief geologist of the Tennessee Survey, for his help with the literature on his state's Crossville Sandstone, and the assistance of Pennsylvania slate expert Joseph Jenkins, who provided tutorials on the identification of stone roofing tiles.

In the second and equally critical area, architecture, a broad range of professionals greatly enhanced my understanding of the Cream City's built environment. Of these, I must first acknowledge Stephen Kelley, noted historic preservation specialist and Visiting Scholar at the University of Wisconsin–Milwaukee. Stephen kindly agreed to read and review my entire manuscript and honored me with an encouraging appraisal. Separately, I obtained a wealth of information about the foundation and building materials of the Northwestern Mutual Tower and Commons from Mig Halpine, director of communications at the New Haven–based firm of Pickard Chilton Architects. And the fascinating story of the birth of the 100 East Building was revealed by three architects who were deeply involved in the design of this signature skyscraper three and a half decades ago: D. Patterson Campbell, now of LS3P Associates; Richard Bartlett, Bartlett Hartley & Mulkey Architects; and Michael Murray, Wagner Murray Architects. I extend my special gratitude to these three preeminent architects for freely sharing their expertise and recollections with me.

Still, however great the assistance of the abovementioned architects and geologists, the input of other professionals—civic officials, museum administrators, and curators, engineers, civic officials, and businesspeople in the building-materials trade—was no less important to this enterprise. It was a signal honor to have this book's manuscript reviewed by Milwaukee mayor Cavalier Johnson and his staff. And I especially appreciate the assistance of Rebecca Ehlers, vice president of marketing, communications, and visitor experience at the Milwaukee Public Museum. Rebecca kindly oversaw her staff's review of the MPM's description, and of other sections dealing with Milwaukee County's bedrock geology. At the Sarah and Charles Allis House, Taytum Markee provided a list of the ornamental stone types on site, and her very informative docent staff treated me to a splendid tour of the museum and its priceless art collection. Later in the writing process, the Allis collections manager Jenille Junco provided crucial additional information and kindly reviewed my manuscript section on her site. At the landmark St.

Paul's Episcopal Church, docent Paul Haubrich gave me a thorough rundown on that landmark's Lake Superior Brownstone exterior and general construction history. And various members of the Forest Home Cemetery office staff helped me locate its geologically significant grave sites and their monuments.

My ever-growing roster of outstandingly helpful stone and ceramic-material producers includes Cleveland Quarries president Zachary Carpenter; Marco Pezzica at Colorado Stone Quarries; Kevin Aune of Ohio's Briar Hill Stone; Sheena Owen of Toronto's North Country Slate; Polycor's Sylvie Beaudoin; Sue Lockwood at Dakota Granite; Coldspring Corporation's marketing coordinator Stacy Gregory; Tennessee Marble Company's Josh Buchanan; and George Hibben of Cincinnati's famous Rookwood Pottery.

Also essential to the research phase of this book were sources and records unearthed by the staff of the Milwaukee Central Public Library and the Milwaukee County Historical Society. At the latter institution, I owe special thanks to Executive Director Ben Barbera, who kindly reviewed this manuscript, and to archivists Kenneth Abing and Steve Schaffer, whose command of the local historical literature was most impressive. In the same vein, I found that local historians in far-flung Wisconsin quarrying towns had much information and perspective to share. These include Amberg Museum curator Ken Jones, Bobbie Erdmann of the Berlin Historical Museum, Kathleen McGwin of the Montello Historic Preservation Society, and former Montello Granite quarry owner Bryan Troost. And, much closer to home, I was able to ascertain the current status of Wauwatosa's Schoonmaker Reef with help from both that city's mayor, Dennis McBride, and David Simpson, the director of public works.

While the body of information supplied by all these individuals, institutions, and companies has done so much to make this book all that it can be, I still am forced to acknowledge, as humbly as this unhumble soul can, that it's only the first tentative telling of a great story on a great theme. As much as I've tried to root them out, sneaky, subterranean mistakes remain. As the petrologist Robert Folk once wrote, "None of the statements herein are to be regarded as final. . . . Such is the penalty of research." I can only hope that other geologists will conduct that additional research, and add to the literature of this great city and subject.

MILWAUKEE IN STONE AND CLAY

INTRODUCTION

The magnificent city of Milwaukee asks something of us, and it's something unexpected. It urges us to love it not only for its vibrant human history, its diverse cultures and neighborhoods, but also for its remarkable geologic legacy. In that matter, too, it claims our curiosity. Not just a notable place for residents to live or tourists to visit, Milwaukee and its surrounding communities are a magic portal that leads us into a greater nonhuman world—a world of immense physical and biological forces acting over eons. A multitude of stories from Earth's past stand ready to be discovered and relived.

Those stories begin deep beneath the skyscrapers of the Cream City's Juneau Town, where there lies a mass of crystalline basement rock well over a billion years old. Sitting atop it are layers of light-gray stone, deposited in a shallow subtropical sea 425 million years ago. It was a time when this region lay south of the equator, and when corals and other marine organisms built complex barrier reefs. Rising higher, there are still other beds composed of sand and clay and silt and boulders, the relicts of the much more recent Pleistocene Ice Age. And perched above all these, in the sunlit world we surface creatures inhabit, stand buildings and monuments adorned in stone, brick, terra-cotta, metals, and other geologically derived materials. They tell still more stories, often of places and times far from our own. These materials come from ancient fragments of the Earth's crust that formed when our planet was only one-quarter of its current age, and from the North African desert, the Apuan Alps of Italy, the Caledonian highlands of Scotland. On the city's streets and in leaf-dappled cemeteries, structures and monuments of rock and fired clay tell their tales of surging glaciers, meltwater lakes, colliding landmasses, disappearing oceans, teeming coal swamps, and continent-rending rifts with unimaginable outpourings of incandescent lava. So, from this familiar landscape of modernity we can be transported to other exotic geographies: long-vanished lands with names like Laurentia, Avalonia, and the Yavapai Terrane. In all directions stretching away from us are revelations of beauty, significance, and awe. Far from being divorced from nature, this urban region is a direct and undiluted expression of it.

This book seeks to tell you these stories and others, and to inspire you to seek out more on your own. In doing so, it delineates the unexpected and often surprising links found here—links between geology and architecture, between the conceptual worlds of the artist, the scientist, and the engineer. And it encourages you to see rather than to not see, to observe what both architects and geologists can when most persons cannot—to understand how all the details become, in one wonderful moment, a higher vision of grand themes and interrelationships finally unveiled.

Part I
FUNDAMENTALS

NOTES ON THE BOOK'S FORMAT, AND TIPS FOR EXPLORING MILWAUKEE COUNTY'S GEOLOGY

This book, the second in a series on the architectural geology of the Badger and Prairie States, employs the same format and approach as its predecessor, *Chicago in Stone and Clay*. It, too, is a bridge between the worlds of science and art, and it's intended to offer the reader and explorer of Milwaukee County a plethora of facts and ideas from the remarkably interconnected worlds of geology, architecture, civil engineering, and history without being overly technical or didactic. I haven't hesitated to inject my personal observations, experiences, and second-childhood enthusiasm along the way. In my decades of writing and teaching I've learned that this approach often humanizes the subject matter in a way that helps the reader who might otherwise feel overwhelmed by both the wealth of detail and the immensity of geologic time discussed.

In fact, there is a long tradition, in Wisconsin and beyond, of writers in science who've understood that their subjects transcended science alone; that they had a responsibility to communicate in an engaging and educational way with the public at large. They knew that what they'd learned and now wanted to describe was too important to be kept as arcane knowledge available only to an academic or social elite. An early example of this mind-set was Milwaukee's own Increase Lapham (1811–1875), an entirely self-taught scholar whose accomplishments in geology, geography, botany, archaeology, meteorology, and science writing have been widely acclaimed. He, and such other naturalist-notables as his German contemporary Alexander von Humboldt and John Muir (another man with strong Wisconsin connections), combined scientific description with personal observations, aesthetics, and an impassioned point of view.

While part II and its descriptions of specific sites make up the bulk of this book, the remaining chapters of part I also play an important role in laying out the geologic context: Milwaukee County's geologic history, and the composition of its bedrock and sediments that figure prominently in the construction of the region's built environment. Then you're introduced to origins and classification of geologically derived building materials—from stone, brick, and terra-cotta to concrete, plaster-based media, and ornamental metals. But first, I offer the following tips based on my own experiences exploring and leading architectural-geology tours.

Site Names

In those cases in Milwaukee County where buildings have more than one name, reflecting changes in owners or occupants, I most usually use the original one cited by the Wisconsin Historical Society or other architectural authorities unless there is some compelling reason for preferring the current moniker instead. In any case, I also list common alternatives in the header of each part II site description.

The Author Is Not Responsible for Any Missing Buildings, Statues, Monuments, or Portions Thereof

In America, cities and towns tend to change their buildings and sometimes their utter souls with a rapidity that baffles Europeans who live in neighborhoods that are centuries old. One example is Milwaukee's Bradley Center. While I personally would never have accused it of being one of the handsomest structures in this architecturally distinguished town, I did eagerly look forward to citing it as an excellent outcropping of the "Carnelian" variety of South Dakota's ancient Milbank Area Granite. On one of my visits it was there; on the next, it was gone. In this book's lifetime, other such disappearances are all too likely. But I have done my best to double-check the continued existence of the sites described before this book goes to print.

What's Emphasized and What's Not

Quite a few of the buildings discussed are made of two or more geologically derived building materials. I here follow my own long-standing rule of emphasizing those whose identity I've certified. Still, there are instances when I feel free to

FIGURE 1.1.   Detail of the exterior cladding of Milwaukee's now-demolished Bradley Center. While it was hardly the most architecturally distinguished building in town, it was the perfect place to examine the "Carnelian" variety of South Dakota's Milbank Area Granite. Given the fact the profit motive is an inestimably greater force in American cities than geological curiosity, such disappearances are likely to continue.

speculate on undocumented materials as long as I make it plain that I am indeed speculating. Happily, there are also some wonderful sites where a larger inventory of materials can be discussed with a high degree of confidence.

Be Respectful of Private Residential Property

The majority of sites discussed in this book, whether publicly owned or not, are open to visitors willing to respect their hours of operation and security regulations. But in some cases the buildings mentioned are private residences. By

FIGURE 1.2.   The "Tudorranean"-style Wiswell House (site 8.21), in Milwaukee's Northpoint neighborhood. It's one of many homes in the region that are both geologically and architecturally significant. But most still serve as family residences, and their owners' privacy and property lines should be respected.

all means enjoy these, but don't jump fences, wander up driveways, loiter on doorsteps, or otherwise trespass to get a better look. If you do, you might just be mistaken for a municipal building inspector or, at your nontrivial peril, someone even more threatening. Ornamental details or materials that must be seen at a distance are best studied through some sort of magnification device. But if you do use one, be discreet. You don't want to appear to be casing the joint without permission.

The Compleat Architectural Geologist: What You'll Need to Bring on Your Explorations, and What You Shouldn't

On some of my tours I've been simultaneously amused and alarmed by normally well-centered participants who appear to be equipped for a three-month safari in the Serengeti. Among the gear I've seen are rock hammers, rock chisels, oversized binoculars, shoulder-mounted video cameras, and multivolume reference libraries stuffed into designer backpacks. Some of these are just too burdensome and distracting; others are completely out of place and could be perceived as lethal weapons by passersby and the local constabulary. The most I'd recommend is a hand lens, perhaps a small pair of field glasses or a camera with zoom capability, and—if it actually needs saying—this book.

On no account should you plan to collect specimens of stone, brick, terracotta, or anything else found on the sites described herein. Leave the hammer and chisel at home, and restrict their use to your wanderings in the desert places of this world. Also, while at a few points in the text I mention stone types that can be identified by their reaction to dilute hydrochloric acid, do not bring your own supply. That also goes for vinegar and other acidic stand-ins. Such substances can leak, stain, stink, and even cause serious injury. And it is not advisable to put any foreign substance on a building, even if it appears to do no damage. If nothing else constrains you, remember that this is the Golden Age of Surveillance.

MILWAUKEE COUNTY'S GEOLOGIC HISTORY AND SETTING

How Geologic Time Is Measured

Geology is a great and stout-trunked tree growing in the magical grove of natural history. In the three centuries it has been recognized as an established science, this tree has put forth a multitude of fruit-bearing branches that extend in many directions and sometimes interweave with the branches of other trees—physics, chemistry, biology. One of the oldest and most productive of these branches is historical geology, which applies discoveries from such fields as paleontology, stratigraphy, petrology, and geochronology to construct a detailed account of the events that have transpired on our planet since its origin about 4.54 billion years ago.

Historical geology is such an important branch because wherever geologists are and whatever geologic features they're looking at, they have a compulsive need to orient themselves in space and time. The questions that drive them are "What does this landscape around me tell me about the way things used to be?"— "What are the rock units and structures in the Earth's crust around and below me, and what do they reveal about the past?" These are questions that could not be adequately answered until the formulation of a detailed chronology, which is handily expressed in the *geologic time scale*.

To use this scale and understand its implications is a humbling experience. The immense totality of geologic time is almost incomprehensible to a mind confined to the everyday concerns of one human life. But if you're willing to let the familiar horizons drop away and you venture forth into the inhuman realm of geologic time, you'll see your own subatomic size set against the vast backdrop of billions of years. For example, if you try to locate our own origin as a species roughly 300,000 years ago on the time scale provided in figure 2.1, you'll find it's invisibly embedded in the ink of its upper line. And even if you were to construct your own half-mile equivalent and lay it in a straight line along downtown Milwaukee's Wisconsin Avenue, from its eastern terminus to the river bridge, humanity's span would only comprise the last two inches.

But the contemplation of the geologic time scale can also inspire and enrich your life, especially if you sample the cornucopia of information it provides in smaller, more digestible portions. Fortunately, geologists are now able to slice up Earth's immense history into such portions using an internationally accepted hierarchy of categories. At the top of this hierarchy come the *eons*, the largest, truly huge subdivisions of the Earth's entire history. None is less than half a billion years long. Then come the eons' major subdivisions, the *eras*, which in turn are composed of *periods*, which are further sliced into *epochs*. But note that geologists use these terms precisely and never interchangeably. A historian of human affairs may refer to the Elizabethan era in one sentence and the Elizabethan period in another; but the Archean eon in Earth science can never be the Archean era, period, or epoch. And in creating a scale involving such a vastness of time geologists also rely on standard abbreviations: "Ga" signifies billions of years, "Ma," millions; and the uncapitalized "ka," thousands. This shorthand will be used throughout the book.

The figure 2.1 time scale shows that each of these spans of time, from eon all the way down to epoch, has been given its own distinctive name. To the Milwaukee County geologist, the most germane are the Silurian and Devonian periods of the Paleozoic era of the Phanerozoic eon, because the bedrock closest to the surface dates to those times. But also of overriding importance is the Pleistocene epoch, when glaciers of our planet's most recent ice age deposited an immense amount of sediments and created southeastern Wisconsin's fascinating landforms, from eskers and drumlins to kettles and pitted outwash.

One of the most exciting aspects of architectural geology is that wherever you study it, the building materials used there introduce you to additional geologic time frames and stories that add to the local ones. In Milwaukee and its surroundings there are, in addition to the Silurian dolostone that adorns a bevy of buildings, Mississippian-subperiod limestone from Indiana, Oligocene-epoch marble from Italy, Minnesota gneiss dating to the far-distant Paleoarchean era, and much more.

Origins

Geologists have long sought to determine the age of the Earth, but happily we live in a time when we're finally quite sure what it is. Earlier estimates ranged

FIGURE 2.1. The geologic time scale with representative Milwaukee County building materials listed by age(s).

EON	ERA	PERIOD	SUBPERIOD	EPOCH	AGES OF SOME MILWAUKEE COUNTY BUILDING MATERIALS
PHANEROZOIC 541 Ma–present	CENOZOIC 66 Ma–present	QUATERNARY 2.6 Ma–present		HOLOCENE 12 ka–present / PLEISTOCENE 2.6 Ma–12 ka	In one or both epochs: Tivoli Travertine; clays for Cream City Brick; Trenton Gravel clays for Philadelphia Brick
		TERTIARY 66–2.6 Ma		PLIOCENE 5–2.6 Ma	
				MIOCENE 23–5 Ma	Carrara Marble (metamorphosed form)
				OLIGOCENE 34–23 Ma	Carrara and Yule Marbles (metamorphosed form)
				EOCENE 56–34 Ma	Chiampo and Pinoso Limestones; Montagnola Senese and Yule Marbles (metamorphosed form)
				PALEOCENE 66–56 Ma	
	MESOZOIC 252–66 Ma	CRETACEOUS 145–66 Ma			Edwards and Hauteville Limestones; Numidian Red Breccia; Larissa Ophicalcite; Raritan Formation clays for Atlantic Terra Cotta
		JURASSIC 201–145 Ma			Alicante, Botticino, and Rosso Ammonitico Veronese Limestones; Furnane and Numidian Red Breccias; Larissa and Levanto Ophicalcites; Carrara and Montagnola Senese Marbles (protolith)
		TRIASSIC 252–201 Ma			Carrara and Montagnola Senese Marbles (protolith)
	PALEOZOIC 541–252 Ma	PERMIAN 299–252 Ma			Larvikite Monzonite; Narragansett Pier and Buddusò Granites; Cottonwood Limestone
		CARBONIFEROUS 359–299 Ma	PENNSYLVANIAN 323–299 Ma		Buddusò Granite; Crossville and Massillon Sandstones; Illinois Basin shales and underclays for Chicago- and St. Louis-based terra-cotta makers, and for St. Louis Brick
			MISSISSIPPIAN 359–323 Ma		Buena Vista Siltstone; Salem Limestone; Yule Marble (protolith)
		DEVONIAN 419–359 Ma			Aberdeenshire, Hallowell, Mount Waldo, Vinalhaven, and Woodbury Granites; Barre and Nictaux Granodiorites; Berea Sandstone; Milwaukee Formation Dolostone; Temiscouata Slate (both protolith and metamorphosed form)
		SILURIAN 444–419 Ma			Somesville Granite; Eramosa, Fond du Lac, Lannon, Lemont-Joliet, Waukesha Area, and Wauwatosa Dolostones
		ORDOVICIAN 485–444 Ma			Potsdam Sandstone; Holston Limestone; Oneota and Prairie du Chien Group Dolostones; Bushkill Slate; Murphy Marble (protolith); New York Red, Vermont Mixed, and Vermont Unfading Green Slates (metamorphosed form)
		CAMBRIAN 541–485 Ma			Potsdam Sandstone; Murphy Marble (protolith); New York Red, Vermont Mixed, and Vermont Unfading Green Slates (protolith)
PROTEROZOIC 2.5 Ga–541 Ma	NEOPROTEROZOIC 1.0 Ga–541 Ma				Imperial Porphyry; Milford and Westwood Granites; Hinckley Sandstone and Lake Superior Brownstones including Chequamegon and Jacobsville Sandstones (probably; all could be Late Mesoproterozoic to Early Cambrian); New York Red, Vermont Mixed, and Vermont Unfading Green Slates (protolith)
	MESOPROTEROZOIC 1.6–1.0 Ga				Mellen Gabbro; Isabella Anorthosite
	PALEOPROTEROZOIC 2.5–1.6 Ga				Athelstane, Montello, St. Cloud Area, and Wausau Granites
ARCHEAN 4.0–2.5 Ga	NEOARCHEAN 2.8–2.5 Ga				Hinsdale and Milbank Area Granites; Lac du Bonnet Quartz Monzonite
	MESOARCHEAN 3.2–2.8 Ga				
	PALEOARCHEAN 3.6–3.2 Ga				Morton Gneiss
	EOARCHEAN 4.0–3.6 Ga				
HADEAN 4.54–4.0 Ga					

from thousands to millions of years, but as the twentieth century rolled on it became increasingly clear that it must be pushed back even farther, well over the one-billion mark. The effort to achieve the date we now consider definitive, and to delineate more recent events in Earth history as well, has required one of the greatest and most sustained examples of scientific sleuthing. It's one that has involved thousands of men and women who've gathered information and formed hypotheses over the course of several centuries. Their patient and often unsung efforts have given us one of the most profound acts of consciousness-raising humankind has ever known. The types of evidence required to succeed in this great undertaking include the mapping and sequencing of rock units in the field; the study of fossils, glacier ice cores, and ocean sediments; and the isotopic analysis of both terrestrial rocks and meteorites formed in the earliest days of the solar system. All of these provide crucial clues that have permitted us to fill in the geologic time scale to a degree that would have astounded earlier generations of researchers. And from these clues we now understand the Earth is approximately 4.54 billion years old—almost exactly one-third the age of the entire cosmos.

The twentieth century also witnessed a revolution in geology in the form of the emerging theory of plate tectonics. The Earth's crust and mantle, once thought to be essentially static, were shown to be remarkably mobile instead. We now have multiple lines of evidence that continents move great distances, collide, grow, and are torn apart by the formation of giant rift zones. Ocean basins expand, shrink, and even disappear as they are subducted deep into the interior over the course of millions of years. As you'll see in part II, the building stones on display in Milwaukee County offer their own evidence of many of these processes and events.

The earliest version of North America, dubbed Laurentia by geologists, formed in the Paleoproterozoic era when a section of Archean crust known as the Superior Craton merged with several other similarly aged fragments and was subsequently bulked up further by the addition of other wandering terranes. (The term *terrane*, not to be confused with *terrain*, is geo-jargon for a substantial section of the crust that has its own origin and identifiable characteristics.) While much of the Superior Craton lies in what is now Canada, it is also exposed in a few places in northern Wisconsin and Michigan's Upper Peninsula, and more extensively in Minnesota. In contrast, southeastern Wisconsin is mostly formed from somewhat younger crustal blocks, the Mazatzal Terrane and the Eastern Granite-Rhyolite Province. Laurentia and its successors Laurussia, Laurasia, and early North America have traveled far in their time; at 650 Ma, near the end of the Proterozoic eon, Laurentia was situated deep in the Southern Hemisphere, not far from the modern location of Antarctica. In the Pennsylvanian subperiod, some 350 Ma later, our portion of what had become the supercontinent Pangaea straddled the equator.

FIGURE 2.2.   The stone cladding of this postmodernist Juneau Town skyscraper (site 5.34) is the Neoarchean-era Lac du Bonnet Quartz Monzonite, quarried in southeastern Manitoba. It formed late in the Algoman mountain-building event in which the Superior Craton, one of North America's oldest geologic provinces, was sutured together. Note the foliation (linear bands of dark minerals). This suggests that this igneous rock was partially metamorphosed at some point in its long history.

One chapter of this long continental saga, the Midcontinent Rift event, is of special interest to Milwaukee's architectural geology, because some of the most striking stone types used here formed either during its vast upwelling of igneous rock, or in its aftermath, when sediments collected in the huge breach it had created in the Earth's crust. The Midcontinent Rift, often abbreviated MCR, occurred approximately 1.1 Ga ago, near the end of the Mesoproterozoic era. This huge, 1,200-mile horseshoe-shaped structure runs from Oklahoma up to the Lake Superior region and then back down to Alabama, though its existence outside the Lake Superior region can usually only be inferred by sophisticated sensing techniques. While geologists still have much to learn, one recent and persuasive interpretation, based on new geophysical data, posits that the MCR is an unusual hybrid feature that formed in a two-step process. First, a passive rift came into being when Laurentia, an ancestral version of North America, separated from another continental plate, Amazonia. Then the rift overrode a mantle

FIGURE 2.3. A hefty chunk of Mellen Gabbro that forms part of John Barlow Hudson's *Compass* sculpture (site 8.14) in Milwaukee's Lower East Side neighborhood. This intrusive igneous rock comes from the vast body of magma that filled and underplated the Midcontinent Rift. The parallel grooves are from holes drilled in the quarry to help split the stone.

plume that triggered a massive outpouring of lava and emplaced intrusive rock below the fragmented, sunken surface. And the MCR also created a small fault-bounded section of the crust and upper mantle between its eastern and western limbs. As disturbing as it may be to most inhabitants of Milwaukee and other Wisconsin communities, this chunk of real estate they're living on is now called, by some geologists at least, the Illinois Microplate.

The Formation of the Region's Bedrock: the Paleozoic Era

While some other portions of Wisconsin are composed of terranes that now expose Proterozoic and even Archean rock at the surface, Milwaukee County's uppermost bedrock dates just to the **Paleozoic era** (541–252 Ma), the first subdivision of our current, Phanerozoic eon. This was the time of the great diversification of multicellular life, and also one episode after another of plate-tectonics activity, climate change, and sea-level fluctuations that repeatedly transformed our continent's shape, weather patterns, and living communities. And it was in the Paleozoic that the plant and animal kingdoms expanded dramatically notwithstanding this era's mass extinctions.

The portion of the Paleozoic most evident in both the uppermost crust and building stones of Milwaukee County is the **Silurian period** (444–419 Ma). And its locally predominant rock type is dolostone, formed from limey mud that was deposited on the floor of a shallow subtropical sea. In this some of the Earth's first coral reef communities flourished. The remains of these complex ecosystems can still be glimpsed in such places as Milwaukee's Soldiers' Home Reef and Wauwatosa's Schoonmaker Reef.

The importance of this region's Silurian dolostone was noted by early Milwaukeean geologist and naturalist Increase Lapham, who in standard nineteenth-century fashion did not clearly distinguish this rock type from the closely related limestone. In the 1846 edition of his groundbreaking survey of Wisconsin's natural features, this self-educated Quaker polymath wrote that "this . . . great deposit of calcareous rock extends south into Illinois, where it dips under the rocks of the 'coal formation,' and north it continues along the lake shore as far as Mackina. It generally occurs in thin, compact layers or strata,

FIGURE 2.4. An old headwall exposing one section of Wauwatosa's Silurian-period Schoonmaker Reef (site 6.33). Rock taken from this former quarry and others nearby, the Wauwatosa Dolostone, can be seen today on some of Milwaukee County's finest nineteenth-century buildings.

disposed horizontally, or nearly so, of a light grey color, and affording very pure lime when burned. At other places it is of a dirty yellow color, filled with minute pores, and easily decomposed, when exposed to the air and weather." He also noted the effect this relatively hard stone had produced on the local landscape, and delineated its economic potential:

> There appears to be a tendency in this rock to assume the form of irreg-
> ular and moderately elevated ridges extending throughout its whole
> length, in a general northeast and southwest direction, or parallel to all
> the geological formations of the western country. It contains dissemi-
> nated masses of sulphuret of zinc, and iron pyrites—the latter in some
> places in great abundance. Cavities are also found filled with bitumen,
> resembling the petroleum or Seneca oil; but there are no localities from
> which this substance can be obtained in any considerable quantities. . . .
> The horizontal layers afford an excellent building material, and would
> answer admirably for pavements, hearth-stones, &c., &c. It is occasion-
> ally found with sufficient firmness of texture to receive a high polish;
> and is then called marble.

FIGURE 2.5.   Devonian-period Milwaukee Formation Dolostone outcropping along the Milwaukee River in Estabrook Park (site 8.29). This richly fossiliferous sedimentary rock was once extensively quarried here for the production of natural hydraulic cement.

Ample mention should also be made of the next, Devonian period (419–359 Ma), because the bedrock of northeastern Milwaukee County dates to that period instead. Mostly a succession of shale and carbonate strata, it has not been used extensively for building stone, though at the locale now occupied by Shorewood's Estabrook Park, dolostone of the Devonian Milwaukee Formation was previously extracted in great quantities for the manufacture of high-quality natural hydraulic cement. The exposed rock faces in the falls area in Estabrook Park provide the geological explorer with a good glimpse at a period well represented across the big water in Michigan's Lower Peninsula, but not otherwise often encountered in the Badger State. And, besides being key to a crucial aspect of the region's architectural development, the Milwaukee Formation offers an amazing look at Devonian animal and plant life contained in its abundant fossil record. Considering both this and our local Silurian reefs, there's no wonder Milwaukee County has been the stomping ground of some of the world's most noted paleontologists.

The Quaternary Period

While the variety of Paleozoic units in our area is an excellent reminder of the local bedrock's considerable antiquity, we should never lose track of the fact that much of southeastern Wisconsin's landscapes, both built and natural, have been shaped by much more recent geologic events of the **Quaternary period** (2.6 Ma–present). This span has contained the dramatic sweep of Pleistocene Ice Age glacial and interglacial phases as well as the Holocene epoch warm-up, perhaps just the latest interglacial prolonged by the climate-altering activities of our own upstart species.

If to the architecturally minded Milwaukeean the Silurian period provided a well-stocked storehouse of useful and attractive dolostone, the Quaternary must certainly be considered an equal gift to humanity. It has provided a seemingly endless abundance of unconsolidated sediments: clays and silts for our brick industries, gravel for foundations and roadbeds, and pebbles, cobbles, and boulders for fieldstone construction and landscaping ornament. Geologists classify these sediments as

- *Till*, an unsorted mixture of all particle sizes from clay to boulders that was laid down directly below or in front of a glacier during one of the later ice sheet advances of the Pleistocene epoch;
- *Outwash*, sands and gravels deposited by flowing streams during periods of glacial melting;
- *Erratics*, detached larger rock fragments carried south by the ice sheets;

MAP 2.1. A map of the Midwest showing the locations of large structural features cited in this book, including the Michigan Basin and the Illinois Basin (© Indiana Geological and Water Survey, Indiana University, Bloomington; reproduced by permission).

- *Lakebed clays and silts* deposited by proglacial lakes and Lake Chicago, Lake Michigan's higher-standing precursor; and
- *Fluvial (stream-transported) sediments* laid down in the floodplains of the Milwaukee, Menomonee, and Kinnickinnic Rivers.

The Structural Setting

Milwaukee County rests on the western edge of a major, downwarped area of the Earth's crust known as the Michigan Basin. Centered on the Wolverine State's Lower Peninsula, this great bowl-like structure began to form in the Cambrian period, some 500 Ma ago, as part of a larger feature that included what is now the Illinois Basin, to the south. It was subsequently separated into its current, smaller extent by the rise of the Kankakee Arch (see map 2.1). The western flank of the Michigan Basin extends into eastern Wisconsin, which accounts for the slight downward slant of our Silurian and Devonian strata to the east.

THE GEOLOGY OF BUILDING MATERIALS

Not a day passes but specimens are brought of new materials, granite from Wisconsin, sand-stones from Michigan, onyx from Mexico, marbles from Colorado to California. There is an equally steady current of new processes for art-metal work in bronze and iron, of mosaics in glass and marble.

—John Wellborn Root, from a posthumous 1892 article published in *Home in City and Country*

Some of the materials architects have chosen for their buildings and monuments cry out for our attention: gloriously patterned marbles, rugged-faced granites, gleaming metallic alloys, glittering inlaid stone. Others, such as unassuming drab-gray limestones, common brick, and concrete, humbly stand in the background of our conscious minds but nevertheless add significantly to the overall visual effect of the structures containing them. Regardless of their showiness or lack of it, all of these materials are to the geologist and urban naturalist a never-ending source of fascination in their own right. Each has its mythic origin story; each speaks of long-vanished environments and natural processes taking place thousands, millions, or even billions of years ago. While lofty mountainsides, steep-walled canyons, and remote rock outcrops certainly have much to teach us about Earth history, it's ironic that it's closer to home, in our cities and towns, where we can derive the most concentrated and diverse lessons in this vast and epic subject.

Building Stone

Most often, building stone is obtained from *quarries*, excavated sites where rock still attached to the Earth's crust is dislodged by any method that works, from raw human muscle power—the ancient Egyptians often just used one heftable piece of rock to pound on another—to sophisticated mechanical techniques and the blasting power of explosives. However it's extracted, it is split or cut into smaller

FIGURE 3.1.   Milwaukee's Wisconsin Avenue, both East and West, offers the urban naturalist an amazing assortment of geologically derived building materials discussed in this chapter. They include the gleaming, glazed terra-cotta of the Plankinton Arcade, at right (site 6.4) and the limestone-and-concrete façade of the high-rising 100 East Building in the background (site 5.12).

sections that can then be sent to a stone dealer's stockyard or to the construction site itself. The rock finds its ultimate function either as squared-off blocks termed *ashlar*, or as larger and thinner panels of *cladding* mounted on exterior or interior walls.

Milwaukee County is also an excellent place to discover that building stone of great beauty and appeal can additionally come from detached pieces of rock, already freed from their point of origin by nature's own forces of erosion: glaciers, streams, and surf action. **Fieldstone**, mostly erratics and smaller rocks collected from glacial deposits, can make a boldly textured and strikingly colorful effect when set in a mortar matrix.

One especially artful application of geologically derived materials is **inlaid stone**, a technique pursued by various cultures at various times, but never more brilliantly than in Italy. It involves the use of a wide assortment of rock types and semiprecious gems, as well as wood and other decorative substances, to produce mosaics of exquisite precision and remarkable detail. In Renaissance and Baroque Italy, the two greatest centers of inlaid-stone production were Rome, where softer rock types were usually chosen, and Florence. That Tuscan city developed the Pietre Dure (hard rocks) style that relied primarily on more resistant materials, such as jasper and lapis lazuli.

Monumental stone is often considered its own special category, even though it's extracted in the same way building stone is, and often in the same quarries. As its name indicates, this is rock specifically dedicated for monuments, whether they're civic showpieces erected to commemorate historical figures or events, such as the Civil War memorials so commonly seen in Wisconsin communities, or the elaborate mausoleums and humbler headstones and markers in grave-yards. Such famous burial grounds as Milwaukee's Forest Home Cemetery are first-rate geologic sites offering the urban explorer a rich array of rock types and superb examples of the stonemason's and sculptor's arts.

The Geological Classification of Stone

> **"What's the use of their having names," the Gnat said, "if they won't answer to them?"**
>
> **"No use to them," said Alice; "but it's useful to the people who name them, I suppose. If not, why do things have names at all?"**
>
> —Lewis Carroll, *Through the Looking Glass*

Let's revisit the arboreal metaphor in chapter 2, in which I likened the science of geology to a great and well-ramified tree. If historical geology as described there is indeed one of its stoutest and most venerable branches, it must also be said that *petrology*, the study of rock types, is a still older and stouter branch that grows even closer to geology's roots. Long before the literal interpretation of religious creation-myth chronologies was questioned by early Earth scientists, quarriers and miners sought to accurately identify, differentiate, and catalog rock based on its appearance and use. For instance, the ancient Egyptians working the quarry district of the Aswan region classified the different stone types extracted there; the local red granite was (among other names) *iner n matj*; the black granodiorite, *iner kem*. Centuries later, the Romans classified many more rock types found at places under their sway. For example, what we now call Larissa Ophicalcite they dubbed *lapis atracius*; the now world-famous Carrara Marble, *marmor lunensis*. And, later yet, the stonemasons of the Italian Renaissance flooded the termino-logical market with a bewildering array of sonorous and evocative descriptors, such as *occhio di pavone bianco, fior di pesco, giallo di Siena, cipollino verde* (White Peacock's Eye, Peach Blossom, Siena Yellow, and Green Little Onion). But with the rise of modern geology a more systematic and widely applicable if less poetic approach to classification came into being. At its most basic, it groups rock types into three broad categories based on their composition and mode of origin: *igneous*, *sedimentary*, and *metamorphic*.

IGNEOUS ROCKS

Derived from *ignis*, the Latin word for fire, this category is reserved for those rocks that formed from the cooling and solidification of *magma* (molten rock underground) or *lava* (molten rock that has reached the Earth's surface before hardening). There are also a few varieties that form from the deposition of volcanic ash spewed into the air by some types of eruptions. But actually they too are derived from a recently molten source.

There are two main criteria that petrologists look for in determining the identity of each igneous rock:

Composition: With a choice of *felsic* (rich in the elements silicon and aluminum, and lighter in both color and weight), *mafic* (rich in iron and magnesium, and heavier and darker), *intermediate* (with composition, weight, and appearance more or less halfway between felsic and mafic), or *ultramafic* (very rich in iron and magnesium, and very dark and heavy)

Environment where it solidified: The choice of *intrusive* (formed from magma, and with larger mineral crystals, visible to the naked eye) or *extrusive* (formed from lava or ash, with smaller, often microscopic crystals)

The identity of each of the igneous building stone varieties found in Milwaukee County's buildings and monuments can be described and visualized using this system. The most frequently encountered types are:

Granite: intrusive, felsic; this category includes both regular and alkali-feldspar granites
Quartz Monzonite: intrusive, felsic
Monzonite: intrusive, intermediate
Granodiorite: intrusive, felsic to intermediate
Trachyandesite-Dacite Porphyry: extrusive or shallow intrusive; intermediate to felsic
Gabbro: intrusive, mafic
Anorthosite: intrusive, felsic

Granite, granodiorite, and quartz monzonite are placed together in the general term *granitoid* for the simple reason that they share similar mineral compositions and origins. Monzonite is sometimes defined as a granitoid, too, though other classification schemes exclude it because it lacks appreciable amounts

FIGURE 3.2. Wisconsin's own Athelstane Granite, used as rock-faced ashlar on the lower façade of Juneau Town's Federal Building (site 5.6). This coarse-grained Paleoproterozoic igneous selection is one of the city's most striking and widely used granitoids.

of the mineral quartz. Regardless, all these types resemble each other so much that their exact petrological identities are sometimes impossible to determine without careful analysis in the lab, or at least without the experienced eye of a professional petrologist. No wonder that both stone producers and architects usually lump them all together as granite and leave it at that! On the other hand, geologists know that the subtle distinctions between granitoid rocks can be important, and if understood can lead to meaningful differences after all. For that reason, I follow the geological convention and, for example, refer to the "Barre Granite" of dealers and stonemasons as the more scientifically correct Barre Granodiorite.

Regardless of whether the igneous building stone is granitoid or not, we can see that knowing the criteria set forth above can help us identify it, at least in a general way. For instance, if we use as the example true granite, the most common of igneous building stones, we'll see that it has individual mineral grains or crystals large enough to distinguish without the need of a hand lens. In addition, it's characteristically lighter in color—frequently pink, red, or a rather pale gray. However, gabbro is dramatically different. It's a very dark gray to black, and it is also heavier per unit volume. Still, being intrusive it has one thing in common with granite: its grains are large enough to see without a hand lens.

A wide variety of igneous building stones have played a major role in architecture since at least the time of the pharaohs. Many types are renowned for their beauty, hardness, durability, and ability to be polished into an almost mirrorlike reflectivity.

SEDIMENTARY ROCKS

A common sight in outcroppings in the southern Wisconsin countryside, these types of stone are also abundantly represented in the architecture of our region. They're divided into two groups. The first, *clastic* sedimentary rocks, form by the accumulation and compaction or cementation of sediments that can range from microscopic specks of clay through silt, sand, and gravel to cobbles and hefty boulders. These rocks are most readily distinguished from one another by their particle size. In contrast, *chemical* sedimentary rocks form from the precipitation out of water of previously dissolved compounds. They're identified by their mineralogical makeup—whether they contain calcite, dolomite, quartz (silica), or other compounds.

The most common sedimentary building stones in Milwaukee County are

Dolostone: chemical, composed of dolomite and often some calcite as well
Limestone: chemical, composed of calcite; includes **travertine** (which often has a pitted or mottled texture) and **biocalcarenite** (made up of tiny fossil fragments)
Sandstone: clastic, composed of sand-sized grains; includes the architectural term **brownstone** (sandstone with a red, maroon, or brown color), and a very hard form known as **orthoquartzite**
Siltstone: clastic, composed of silt-sized grains
Breccia: clastic, composed of angular pebbles, cobbles, or boulders; includes **Ophicalcite** (a breccia containing fragments of metamorphic serpentinite and marble)

Sedimentary rocks are also notable because they often contain fossils and such other interesting depositional features as *ripple marks* (ripples in sediments made by wind or flowing water and later turned to stone) and *crossbedding* (a pattern of thin, slanting, and sometimes curved layers in rock, caused by wind or water depositing sediments on a sloping surface like a ripple or dune face).

METAMORPHIC ROCKS

Often notable for their flamboyant appearance and exotic origins, metamorphic stone types are also well represented in our local architecture. As their name implies, these have undergone a profound change of one kind or another—a metamorphosis wrought when the original rock was subjected to immense pressure, searing temperature, or hot, chemically reactive fluids circulating underground. For this category, geologists rely on a variety of criteria to classify the different varieties, but two are most helpful in identifying building stone:

- Whether the rock is *foliated* (with minerals in parallel alignment, producing a banded, wavy, or thinly laminated appearance) or *nonfoliated* (without such parallel patterning)
- Its *protolith* or *parent rock* (which could have been igneous, sedimentary, or even another metamorphic rock)

The most frequently encountered metamorphic building and monumental stones in Milwaukee County are

Marble: nonfoliated; the parent rock is either limestone or dolostone
Serpentinite: nonfoliated to weakly foliated; the parent rock is dunite or other forms of peridotite
Gneiss: foliated; the parent rock is granite or other igneous, sedimentary, or metamorphic rocks
Slate: foliated; the parent rock is shale
Migmatite: a foliated "mixed rock" containing both gneiss or another metamorphosed rock and younger, unmetamorphosed granitoids

Metamorphic rocks range from the tough, hard, and enduring (gneisses and slates) to those that, for all their appeal, are made of minerals much more susceptible to weathering in the unforgiving extremes of our climate (marbles, serpentinites). While some true marbles hold up fairly well when exposed to the elements, their type is usually best reserved for interior spaces.

The Architectural Classification of Stone

> **This has given me the greatest trouble and still does: to realize that what things are called is incomparably more important than what they are.**
>
> —Friedrich Nietzsche, *The Joyous Science*

Geologists would be all too happy if everyone else agreed on their time-honored, rational, and consistent system of names for building and monumental stones. But when we enter the world of architects, builders, quarriers, and salespeople, much of that carefully articulated nomenclature flies out the window. We're in a strange new land with a strange new lingo, and if we are to make any sense out of what we read in historical accounts or at stone-dealer websites, we'd better learn to speak it quickly. It's a language humming with hyperbolic lunacy, marching out badly overblown modifiers, and chock-full of false claims. It's a language in which the clastic rock sandstone is routinely confused with the chemical rock limestone; marble is any softish stone regardless of composition that can take a good polish; and in which mafic gabbro, black as a moonless night, is equated with light-toned, felsic granite. Appearances and marketing razzmatazz matter more than the scientific reality.

Because this book is written from the geologist's perspective, I anchor its terminology in the scientific system whenever possible, and I strive to ensure the stone's rock type is correctly indicated. So I employ a stone's official stratigraphic or unit name when one has been assigned and is still accurate. But if an official name is lacking, I usually identify the stone by its place of origin. In the case of foreign stone selections that are not stratigraphically designated I adopt the most reasonable (or, at any rate, the least silly) version of its trade name, especially if it refers to the stone's locale of origin. These preferred names of mine are capitalized, as in Wauwatosa Dolostone and Montello Granite. However, because the architectural terms for stone are deeply entrenched in the literature, I frequently also give a sampling of each variety's common trade names, too. They're capitalized and set in quotation marks. Accordingly, when I discuss the region's most widely used building-stone type, the Salem Limestone, I often add that it's better known in the building trades as "Bedford Stone" and "Indiana Limestone."

One special situation involves the Silurian dolostone that is the bedrock for much of the area covered in this book. Widely employed in Milwaukee architecture since the mid-1800s, it has been quarried for that application in many places, from the Door Peninsula to Joliet, and even far to the east, in the Niagara Falls region of western New York. While on this side of Lake Michigan it's mostly taken from the Racine and Waukesha Formations in Wisconsin and corresponding units in northeastern Illinois, I've chosen to give the reader a better sense of where exactly each site's stone was produced. I do so by parsing, whenever possible, my general **Regional Silurian Dolostone** category into several more specific varieties. Listed from south to north and named for their towns of origin, they are the

Lemont-Joliet Dolostone (from the Lower Des Plaines River Valley of
   northeastern Illinois)
Wauwatosa Dolostone
Waukesha Area Dolostone
Lannon Dolostone
Fond du Lac Dolostone

FIGURE 3.3. One of two large camel statues that guard the entrance of the Tripoli Temple Shrine (site 6.28), in Milwaukee's Concordia neighborhood. The stone used here, Indiana's Salem Limestone, is a carbonate sedimentary rock famed for its ability to take fine carving. However, close examination of this magnificent critter reveals some deterioration due to weathering. This inevitable natural process has resulted in some of the rock's fossils becoming more fully exposed at the surface.

While it can be difficult to coax it from the Earth's crust and prepare for architectural or monumental use, not to mention expensive to haul to distant locales, stone has been the monarch of building materials for millennia. No other material can quite match its sense of solidity, permanence, and grandeur. And however impressive or disappointing a statue or a structure is in its overall aspect, the stone composing it is often beautiful in its own right. For this reason builders and sculptors have for untold generations carefully studied and skillfully employed the ornamental attributes of marbles, granites, and many other rock types.

Preparing Stone for Architectural Use

In its natural setting, rock exposed to the elements soon forms a weathered surface. To see its true unaltered appearance, the geologist wielding a soft-steel hammer cracks off a piece from its outcrop to see what lies beneath, where still unblemished mineral crystals reveal the rock's true color and texture. But stonemasons have long known that they can cleverly tinker with the stone's natural appearance—its tone, reflectivity, and texture—to achieve various ornamental effects.

This process of transformation is termed *finishing*. First, a piece of rock is taken from the headwall of its quarry. Then it's split, sawn, or otherwise shaped into the preferred dimensions. Either at the quarry site or at another facility it undergoes further processing. Abrasives may then be applied to its outward-facing side to produce either a nonreflective honed or reflective polished surface. Also, a pleasing ornamental effect may be achieved with sandblasting or the application of a water jet, an acid wash, or a high-temperature flame. Or the surface can be scored with metal brushes, hammering tools, or chisels. All these techniques give the architect a wide assortment of effects to choose from. Some of these have proven to be quite indicative of a particular style or era. For instance, when the outer surface of a stone block is intentionally chiseled only on its margin and the central portion is left rough, jagged, and projecting, the stone is called *rock-* or *quarry-faced*. Buildings sporting this massive, medieval look were especially in favor during the Richardsonian Romanesque period of the last decades of the nineteenth century.

How Stone Weathers

Building and monumental stone have a reputation for being endlessly enduring, but in fact their different types vary widely in chemical composition and suffer the ultimately unstoppable degradational process of weathering at dramatically different rates. In Milwaukee's Forest Home Cemetery, for example, marble headstones and monuments dating from the nineteenth century have deteriorated to the point that their inscriptions are badly faded and their other once-crisp details appear to be melting away. But their counterparts made of granitoid rock seem completely unscathed after just as many decades of exposure. As this demonstrates, true marbles, for all their popularity and appeal, can be a risky choice for outdoor settings given the Milwaukee region's annual and daily temperature extremes and abundant moisture. The natural carbonic acid contained in even the most unpolluted rainwater reacts with the calcium carbonate in the marble and can ruin its once-chaste surfaces. Fortunately, though, marble's ancient role as a primary architectural stone is still assured when it's utilized in building interiors, where its beauty and luster can be both appreciated and more easily preserved.

Generally speaking, both felsic and mafic igneous rocks, as well as the metamorphic gneisses derived from them, share marble's ability to take a beautiful finish but are much more resistant to weathering, especially when flamed or polished. In contrast, such sedimentary selections as sandstone, limestone, and dolostone can be more problematic. The Chequamegon and Jacobsville Sandstones, the "Lake Superior Brownstone" varieties so much in favor in Midwestern cities in the later nineteenth century, tend to spall and chip, especially when mounted perpendicular to the original bedding plane or not properly dried out before use. And limestone and dolostone both contain greater or lesser amounts of calcium carbonate, which can lead to the same sad fate as marble in exposed locations. However, there are exceptions to these general traits. For instance, the Salem Limestone, that renowned and ubiquitous product of southern Indiana, often displays an admirable inertness in the face of air pollution and climatic extremes. But even it must be situated cannily and relegated to the right applications. If placed at or near ground level, especially next to a sidewalk or roadway, it wicks up water and dissolved deicing salts. This results either in *exfoliation*, where sections of the stone peel away, or *efflorescence*, a white crust accumulating on the surface of the stone. Exfoliation has sometimes also been a serious problem with the various types of Regional Silurian Dolostone so commonly seen in Milwaukee County.

Brick

Most other geology guides to sites of architectural significance discuss only stone and leave it at that. But it's long been my contention that brick—in common with the other materials listed in subsequent sections—is in its own way just as geologically derived a material as rock itself. After all, clay, brick's basic ingredient, is a substance composed of microscopic mineral grains derived from the natural weathering and erosion of stone. And when left to its own devices it's an essential raw material for both life-sustaining soils and the formation of new sedimentary strata.

In essence, bricks are units of dried or fired clay (and sometimes other substances as well) that can be stacked for easy storage and transport. And like building stone, they've had a very long history. Even in locales where suitable rock has been available, they've been a very successful complement or outright competitor. The earliest recorded use of adobe, which is essentially sundried brick, dates from the interval 8,000–10,000 BCE; and evidence of the much more enduring fired brick comes from 3,500–5,000 BCE. The ancient Mesopotamians perfected the practice of molding and glazing ornamental bricks, and centuries later the Romans started using brick architecturally, albeit sporadically, in the reign of Augustus, though these early bricks tended to be recycled clay tiles sawn into shape. Later, Imperial architects systematized brick production and relied on a variety of standard sizes of mass-produced fired bricks to construct their structures' arches and to face concrete walls.

By the nineteenth century, most bricks produced in southeastern Wisconsin were made directly from clay dug from the region's riverbanks, thick deposits of glacial till, or lakebed deposits. To the south, in the major structural feature known as the Illinois Basin, immense quantities of clay were mined from shales (fine-grained sedimentary rock) and underclays (*paleosols*, or ancient soils) associated with Pennsylvanian-subperiod coal deposits. But whatever its source, the clay was either molded by hand in wooden forms or run through pressing machines, then dried and placed in kilns for firing. Some of America's greatest cities—Philadelphia, Baltimore, Chicago, Milwaukee, and St. Louis—became major production and distribution centers of brick, but even small Wisconsin farm towns often had their own clay pits and brickmakers. Usually these were first-generation industries that sprang up out of dire necessity soon after the earliest settlers arrived.

The post-firing colors of bricks made in our region can provide important clues about the composition of the Quaternary sediments from which their clays were mined. In many places in our region, the unusually high calcium and magnesium content of the till and waterborne sediments, both originally derived from our Silurian dolostone bedrock found in easternmost Wisconsin, masks the iron compounds also present. The result is the famous pale yellowish tint that made Milwaukee's Cream City Brick so striking and so sought-after, and which is also characteristic of brick made in other places close to the Lake Michigan littoral, and even in some more inland locations. Conversely, in places where the extra dose of calcium and magnesium is absent, the iron when oxidized in the firing process expresses itself more fully in hues ranging from salmon and rose to the classic brick-red. One can actually draw a geochemical map of the substrates of southeastern Wisconsin and northeastern Illinois by simply determining the coloration of each town's locally produced brick.

While brick has never been assigned official geologic names the way naturally occurring stone has, it still can be classified and identified in several reasonable and helpful ways. One of the simplest is by its quality, the higher grade being *facing brick* and the lower *common brick*. The former term is reserved for the finest, most attractive, and often most enduring types suitable for façades and other places where their ornamental assets can be proudly displayed. The latter, less carefully or less uniformly made, are usually softer, more pebbly, or more porous. Common brick was often made from unscreened glacial till, whereas facing varieties were fabricated from more well-sorted fluvial (stream) deposits or from the coalfield shales and underclays alluded to above. Traditionally, common brick has been restricted to more utilitarian and less visible applications:

buildings' side and back elevations, or basements and other interior walls. However, in recent decades tastes have definitely changed. Now it's eagerly sought out and even purloined at demolition sites for resale as a Retro Chic ornamental building material.

Another handy way of identifying bricks is by their dimensions and shape. In the United States, standard brick size is often defined as 3 5/8 x 2 1/4 x 8 inches. However, two popular nonstandard brick sizes found in our area are

> **Roman Brick**, a facing variety that owes its name to the fact it resembles the long, deep, and vertically thin brick type favored by ancient Roman builders. Its dimensions may vary a bit, but in modern times they're most frequently given as 3 5/8 x 1 5/8 x 11 5/8 inches. That most famous of Wisconsin architects, Frank Lloyd Wright, opted for Roman Brick in some of his most famous designs. In this he may possibly have been influenced by his great mentor, Louis Sullivan, who also used this type to stunning effect.
>
> **Norman Brick**, which can easily be mistaken for Roman, being just slightly less thin vertically. Its dimensions are often listed as 3 5/8 x 2 1/4 x 11 5/8 inches. It too produces a greater sense of horizontality than can be achieved with normal brick.

Brick can also be classified by its locale of manufacture. The types discussed in this book are

> **Cream City** (manufactured till the 1920s in Milwaukee County; its striking cream to pale-yellow color—not the state's equally famous dairy production—gave the city its nickname; manufactured in both common and facing forms)
>
> **Minnesota** (made in some undisclosed part of that state; ocher-colored)
>
> **Ohio** (made in some undisclosed part of the Buckeye State; mauve to brown)
>
> **Philadelphia** (made in Philadelphia, Pennsylvania; one of the most prestigious red facing bricks in the 1800s)
>
> **St. Louis** (from St. Louis, Missouri; hard and very high-quality facing brick; usually red, but the shading varies from orange and pink through bright cherry or brick-red to brown and maroon with iron flecking)

The remarkable ubiquity of brick all too often results in its being taken for granted. But we should never forget its essential role in Wisconsin architecture: the spade that excavated a community's first clay pit was rarely far behind the ax and plow. Nineteenth-century regional brickyards provided early architects with what was essentially convenient, pre-shaped units of metamorphic rock made of naturally occurring sediments artificially lithified by the intense heat of the kiln,

FIGURE 3.4.   Milwaukee County is famous for its distinctive, locally manufactured Cream City Brick. Juneau Town's German-English Academy, site 5.35, displays it to good effect in one version of what I call the Classic Milwaukee Formula. Here the pale-yellow brick sits atop a plinth of white to pale-gray Wauwatosa Dolostone. Also note the intricate terra-cotta ornamentation that complements the other two classes of building materials.

which can reach more than 2,000 degrees Fahrenheit—a process not dissimilar to how some granites and limestones are cooked into gneisses and marbles in the Earth's interior.

Besides being relatively inexpensive and easily shipped, brick is much more fireproof than most types of stone and metal. And in the hands of talented designers and masons it possesses downright dazzling ornamental attributes. The artist and the artisan can play with so many options. In addition to capitalizing on brick's wide array of colors, sizes, shapes, and textures, they can create amazingly elaborate patterns by varying how bricks are set in their *courses* (rows). These can be composed of bricks laid horizontally as *stretchers* or *rowlock stretchers*, vertically as *sailors* or *soldiers*, on end as *headers* or *rowlocks*—or any combination thereof. The wide variety of designs so produced seems almost endless. One of the better-known ornamental patterns is the **Flemish Bond**, where each course is composed of alternating stretchers and headers, with one course's headers centered over the adjoining courses' stretchers.

Finally, texture can be an identification tool. For example, **Rock-Faced Brick** resembles stone ashlar with a rock-faced finish; its outward side has a raised, bumpy, or irregularly projecting surface.

Terra-Cotta

Found in everything from flowerpots to roofing tiles, terra-cotta, literally "cooked earth" in Italian, has the same main ingredient as brick's—clay. In architecture it attains its most glorious forms as glazed and custom-molded building cladding and ornament, and as *faience*, glazed or enameled ceramic tile that can be applied to a surface in one basic color or in intricate polychromatic patterns.

While by the latter half of the nineteenth century Milwaukee had become famous for its distinctive and highly prized cream brick, its larger urban rival about 80 miles down the coast became equally renowned as a center of terra-cotta production. Among the firms producing it there were the Chicago Terra Cotta Company, the city's first, and the Northwestern Terra Cotta Company, located along the North Branch of the Chicago River. Later, the American Terra Cotta Company, situated in McHenry County, Illinois, rose to prominence, as did the Winkle Terra Cotta Company in St. Louis, and Cincinnati's great faience and pottery specialist, the Rookwood Company. Far to the east, New Jersey's Atlantic Terra Cotta Company also successfully competed in the Midwestern market. The handsome productions of all these firms can still be found in Milwaukee County.

The classification of terra-cotta can be done by its basic properties (unglazed or glazed), or by its type of application (roofing tiles, cladding, or faience). And it can be identified by the firm that made it, if known. In this book, terra-cotta produced by the companies cited just above is indicated wherever the historical record permits.

Earlier generations of Wisconsin architects were by no means slow to comprehend terra-cotta's manifold virtues. Its relatively light weight made it less expensive to transport than building stone. And, as a burnt-clay product, it was resistant to high temperatures and flame. By sheathing their buildings' iron or steel structural elements in terra-cotta, developers could more easily comply with new fireproofing regulations, imposed by insurance companies and city codes alike after devastating conflagrations in Chicago (1871 and 1874), Boston (1872), and Baltimore (1904). Then again, from an aesthetic standpoint terra-cotta proved itself worthy of comparison to any other ornamental building material. Not only could it be custom-cast into remarkably intricate patterns suiting the architect's whimsy; it could even mimic choice building stone so closely as to be practically indistinguishable from it. One is obligated to carefully scrutinize the gleaming wall of a bank or post office interior to tell whether it's really choicest Carrara Marble or its ceramic stand-in. And granite, with its mosaic of multicolored crystal grains, came to be copied with amazing precision with the development of the ingenious "Pulsichrome" technique of terra-cotta patterning. Even a professional geologist is all too likely to be deluded by this would-be igneous rock unless there's a telltale crack or chipped corner that exposes the *bisque*, the yellowish baked clay under only a fraction of a millimeter of glaze.

Cement and Concrete

While the former preeminence of the Milwaukee metro area as a source of building stone and brick is still manifest today in many of its architectural landmarks, few modern residents of the region realize that it was also a world-class producer of yet another essential building-material ingredient: *natural hydraulic cement* derived from the Devonian Milwaukee Formation Dolostone quarried in what is now Shorewood's Estabrook Park and adjoining land. Essentially a forerunner of the Portland cement that is predominant today, this substance was blessed with a chemical composition that enabled it to set even underwater. Produced by the Milwaukee Cement Company and its local competitors, it was sold far and wide in the American Midwest, and no doubt contributed powerfully to the rise of its great cities and their bridges and buildings.

Of course, cement's primary role has always been as the major constituent of concrete. If, as noted before, brick is all too easy to ignore, concrete, so omnipresent and utilitarian, is positively invisible to us most of the time. Still, this everyday substance has a fascinating history of human development and use, and it boasts remarkable chemical properties. In modern times it's most often a blend of lime- and clay-containing Portland cement and aggregate that, when mixed with water, sets into stonelike form in an exothermic (heat-releasing) reaction. Nowadays concrete shapes our world to a degree no other building material does. We walk on it, we drive on it, we fill caissons and erect walls with it, and now we even prefabricate and modularize it. The ancient Romans mastered its applications for everything from aqueducts to bridges, and came to depend on it for their greatest structures and most awe-inspiring public buildings, including the incomparable concrete-domed Pantheon. However, their preferred formula, very likely the most enduring ever devised, was made to a recipe considerably different from our own. For one thing, it relied on the volcanic ash called *pozzolana*. Even so, archaeologists have ascertained that concrete was developed much farther back in history; the Romans perfected their version of it, but definitely didn't invent it.

Today, however, we realize our rampant use of the modern formulation of concrete does come at a grave cost. The production of its key ingredient, cement, accounts for 8 percent of our global carbon dioxide emissions—not as much as generated by electricity generation or our vehicles, certainly, but still a matter of grave concern in an age of unchecked global warming. Major concrete producers are now reportedly grappling with this problem.

Plaster-Based Materials

Plaster, too, has an ancient pedigree and is certifiably geologic. While its exact composition may vary, it always incorporates a chemical compound serving as its binder. Traditionally, this was either the mineral gypsum (calcium sulfate) or lime (calcium oxide and hydroxide) derived from the mineral calcite, otherwise known as calcium carbonate. Lime-containing cement, including Portland cement, has also been employed as a binder. In addition, plaster requires water as well as sand or some other aggregate.

Stucco, a common sight in Wisconsin architecture, is plaster utilized to coat or ornament a building's exterior. On the other hand, when plaster is used as the binding matrix for stone, brick, or terra-cotta, it's **mortar**, which sometimes can be quite decorative on its own and can add significantly to a building's visual appeal. And when plaster is artfully blended, tinted, and finished to simulate polished marble or other fancy stone types, it's **scagliola** (pronounced scal-YO-lah). Like stone-simulating terra-cotta, scagliola can be very hard to distinguish from the rock it so convincingly copies. Then again, there's **sgraffito**, the application of tinted layers of plaster that are scraped or scratched in certain places to reveal underlying colors and so achieve a sculptural or other decorative pattern on a wall.

Metal Materials

Many buildings in southeastern Wisconsin are supported, covered, or adorned with an array of metals that are every bit as geologically derived as any other material described in this chapter. They first formed as ores or native elements in the Earth's crust in various ways. Nowadays iron is almost exclusively extracted from Proterozoic or Archean sedimentary rocks, including those of northern Minnesota and Michigan's Upper Peninsula. Other metal-containing compounds

FIGURE 3.5.   A symphony in metal and stone on the façade of Kilbourn Town's Wisconsin Tower (site 6.11). This exemplar of what I call the Grand Art Deco Formula sports an imposing entranceway ornamented with an elegant cast-iron grille, framed by the most ancient and strikingly patterned of all architectural stone, Minnesota's Morton Gneiss. Above it rise vertical bands of the much more sedate Salem Limestone. It in turn is punctuated with cast-iron window spandrels.

accumulated in stream and wetland deposits, or were emplaced by mineral-rich, hot-water solutions infiltrating preexisting rock units. The following metals can be found on our region's roofs, building façades, external and internal ornament, structural support, and sculpture:

Aluminum (including the variant **Coated Aluminum**)
Copper
Bronze (an alloy of copper and tin or arsenic)

Cast Iron
Wrought Iron
Enameled Steel
Stainless Steel
Weathering Steel

One other metallic building material that has played a huge role in our region's architecture and civil engineering is generally not on view. Nestled within its sheath of concrete, **rebar** (a contraction of "reinforcing bar") is generally seen only during a building's construction phase, or years later, when it has corroded, expanded, and become exposed as the cracked concrete around it spalls away. Most rebar is made of rust-prone carbon steel, though more enduring and considerably more expensive types are also substituted in some projects. The purpose of using rebar, regardless of its composition, is to compensate for concrete's Achilles' heel. While it has impressive compressive strength (resistance to crushing), its tensile strength (resistance to being pulled, bent, or twisted) is considerably less when not reinforced.

FOUNDATIONS

The Roots of Milwaukee's Big Buildings

A Tale of Two Soggy Sisters: Comparing the Substrates of Lake Michigan's Greatest Cities

> **Chicago and Milwaukee . . . have grown together and exerted a pull, more or less strong, on all of the land between them.**
>
> —Robert Bruegmann, *Sprawl: A Compact History*

While Milwaukee and Chicago are certainly both centers of world-class architecture in their own right, they also are, like everything else, part of a larger context. That's especially true when we look at both their subsurface geology and the civil engineering practices that have been used to keep their larger buildings stable and secure in an unforgiving medium. In spite of their cultural, economic, and sports rivalries, the Cream and Windy Cities are best understood when compared and contrasted to each other.

Examples of their common origins abound. To the geographer and geologist they're twin siblings—not identical, of course, but surely fraternal and born of the same parent conditions. Both Milwaukee and Chicago began as settlements on the Lake Michigan shore, at east-facing river mouths near the confluence of two or more streams. And both have downtowns largely raised on low wetland ground that in turn was fronted by sand spits and dunes. This is not to deny some differences between them. Milwaukee's wetland zone was bordered by high forested bluffs; Chicago was and still is almost as uniformly flat as a chessboard, excepting a series of fairly subtle ridges that formed as spits during episodes when the lake level was higher. And Milwaukee's rivers debouched into a well-defined bay instead of meeting a bit farther inland. Regardless, in the eyes of aboriginal peoples and early American settlers, both were similarly appealing places: originally, and for untold centuries, as a source of abundant game and plant resources; then much later as major ports that could link the abundance of the agricultural Midwest with the financial assets and the commercial centers of the Northeast.

But these two city sites also shared and still share an Achilles' heel, at least for those wishing to erect really large and heavy structures. The bedrock, always the best anchor for skyscrapers and other massive buildings, is a *very* long way from the surface—roughly 85 to 100 feet down in the Windy City's Loop and River North, and about 50 to 100 feet deeper than that in the Cream City's Kilbourn and Juneau Towns and Historic Third Ward. And what lies above the bedrock in both cases is a succession of oozy, shifting, and potentially treacherous sediments that have been compared to the layers of a rum-soaked jelly cake. These sediments were deposited by a complicated sequence of events during and after the Pleistocene Ice Age. First, there were the advances of the Wisconsin ice sheet spreading *till*, a mixture of all particle sizes from clay to boulders. Then the inundations of Lake Chicago and other predecessors of modern Lake Michigan contributed *lacustrine deposits* in the form of silt and clay. Finally, the flow and periodic flooding of the Chicago, Milwaukee, Menomonee, and Kinnickinnic Rivers added still more *alluvium*—gravel, sand, and clay. And the high *water table* (upper limit of the groundwater) in those deposits near the lake seemed to make the very idea of skyscraper construction in either city overtly foolhardy.

Regional Foundations, and How They Evolved

That certain unwilling materials have been thrust into certain specific places does not make architecture.

—Louis Sullivan, "Reality in Architecture," in *The Inland Architect*, September 1900

As preposterous as it may still seem to take some of the squashiest and gloppiest real estate in America and turn it into showplaces for very tall and very heavy buildings, the histories of both Milwaukee and Chicago have shown that when there's a will, there's a way.

At first, less-than-successful foundation designs were tried. One involved the construction of a shallow and continuous *raft foundation* employing a concrete pad the size of a building's entire footprint. When this was done for Chicago's second Federal Building, completed in 1880, it wasn't long before the raft began to settle unevenly and break. Office workers within the heavy, stone-faced edifice had to contend with everything from peeling plaster and water-main breaks to leaning walls and unwanted drafts of cold winter air leaking through cracks in the masonry. By the 1890s the massive structure was deemed so unsafe that it was scheduled for demolition. Ironically, much of that ill-fated edifice's ornamental and dimension stone was salvaged for use in the construction of Milwaukee's St. Josaphat Basilica, where it can be seen in a much happier situation today (see site 7.4).

Another method, which enjoyed at least partial success especially in Chicago, was the use of employing isolated spread footings that relied on a network of much smaller, individual pads spaced to distribute the structure's weight as equitably as possible. This was the technique advocated by the influential architect Frederick Baumann (1826–1921). Originally the individual pads supported pyramids of dimension stone and rubble from which a building's structural piers rose. Unfortunately, they were broad and bulky, and usurped much of a structure's basement volume—space that increasingly was needed for boilers and machinery. This problem inspired flatter and lighter *grillage* pads consisting of crisscrossing sets of horizontally laid rails or I-beams embedded in concrete. Regardless of the design of the isolated footings, however, substantial and sometimes differential settling did often occur, as happened with such Windy City landmarks as the Monadnock and Old Colony Buildings. Architects compensated by setting building entrances considerably above grade so they would not eventually sink beneath it. But even the most carefully engineered edifice could still end up with tipsy staircases and walls and floors out of plumb.

One other method, that of the *inverted-arch foundation*, was also tried, at least in Milwaukee. In this approach, heavy exterior walls were supported by piers of stone ashlar connected and braced with brick formed into upside-down circular arches. But ultimately the best solution for anchoring the increasingly large and heavy structures of the nineteenth and twentieth centuries came in two forms. And in both of these, the key to success lay in going deep. The first, the *pile foundation*, was a method that originally involved driving long poles called piles into stiffer, lower layers of sediments. It was by no means a new idea. For instance, the architects of medieval Venice were adept at using piles, often clustered together into bundles called *palafitte*, to support the grander churches and palaces of their lagoonal city-state in its own queasy mixture of waterlogged sediments. In the nineteenth and the early twentieth centuries, piles were still made of the traditional material, timber—often the straight-boled trunks of pines and other conifers. In more recent times, steel piles, usually driven all the way to bedrock, have taken their place.

FIGURE 4.1. Holes in the ground can be positively fascinating, as is the case here, at the construction site of Juneau Town's Couture high-rise (site 5.13) as it appeared in early 2022. Just to the right of the large crane a driving rig is noisily pounding a steel pile to bedrock well over 100 feet down. That pile, and 191 others, constitute a modern take on Milwaukee's most frequently used type of large-building foundation.

Piles do have two potential problems, though. If made of wood, it's essential that they're kept in an anoxic, thoroughly waterlogged environment lest they rot. This has indeed been a problem in recent years in Milwaukee because the water table downtown has dropped significantly. (The experts are still arguing about what factor is primarily responsible for this.) And as the upper portions of timber piles under older buildings have become exposed to air, they have in some cases deteriorated badly, necessitating reinforcement, replacement, or even building demolition. And while the introduction of steel has solved the rotting problem, one other issue remains. Driving piles of whatever kind for new building construction creates considerable noise and bone-jarring vibrations that are at least annoying to those in the neighborhood and at times downright injurious to the foundations and walls of buildings nearby.

The second favored method is the *caisson foundation*. This technique had long been used in nineteenth-century American bridge construction, but it seems that its first application to an urban office building was in 1890 for City Hall in Kansas City, Missouri. Nevertheless, it was in Chicago that this technique was fully proven and perfected. This is in large part due to William Sooy Smith (1830–1916), a US Military Academy graduate who, after an early stint as a civil engineer, went on to serve in the Union army as a brigadier general and cavalry

FIGURE 4.2.   Situated just to the east of the Wisconsin River, the First National Bank Building (site 5.19) was designed by D. H. Burnham & Company. This famous Chicago-based architectural firm here opted for rock caissons, a foundation system developed and perfected in the Windy City.

commander during the Civil War. Later, he tried his hand as a gentleman farmer and then finally decided to reactivate his prewar engineering career. In doing so he became one of America's most sought-after designers of bridge and building foundations. Smith first employed the game-changing caisson method to anchor one portion of the famous Chicago Stock Exchange in 1894; after that, his Windy City projects were many.

As employed in Chicago and Milwaukee, caissons are deep and watertight shafts, traditionally 6 or 7 feet in diameter, dug to a stiff, deep layer of till and thus dubbed *hardpan caissons*, or to bedrock, as *rock caissons*. Lined with hardwood lagging secured with iron rings that was removed only when the concrete was finally poured in, the shafts were also often flared out at the bottom into a conical "bell," especially if they did not reach down all the way to the bedrock. This widened terminus provided an especially solid footing. Until the middle of the twentieth century all of the many caissons needed at each new construction site were laboriously excavated by hand, usually with three-man crews consisting of the supervising signal man or headman, the dumper, and the hand miner in the shaft itself. The crew's progress downward was measured in *sets*—units of depth corresponding to the height of the slats used to line the caisson's wall. Normally the slats were 5 feet, 4 inches long, but shorter ones of as little as 3 feet, 6 inches were used if the substrate was especially soggy. Each caisson team was reportedly expected to dig three sets per day for a total of 16 feet, but George Manierre, writing in 1916, noted that in Chicago 11 feet of progress per eight-hour shift was typical in softer clay, and only 5 feet in hardpan. Starting in the 1950s, however, this backbreaking work was replaced by machine-digging methods.

In contradistinction to Chicago, most Milwaukee builders have opted for piles rather than caissons, though examples of the latter are not lacking in the Cream City, especially at very groundwater-saturated sites near the bank of the Milwaukee River and where the percussive effects of pile-driving simply can't be tolerated. And ironically Milwaukee is now also a showplace of the latter-day resurrection of the shallow raft method, used in particularly waterlogged sites. Taking advantage of amazing technological advances undreamed of a century and a half ago, a method once discarded in favor of deep stabilization can now be effectively utilized. A thick but shallowly set steel-reinforced concrete pad extends under the structure's entire footprint.

Regardless of the method used, it's fitting that we honor the all-too-unappreciated genius, expertise, and hard work of the civil engineers and foundation contractors who've made the great buildings of the Milwaukee and Chicago skylines so secure in their unstable substrates. They and their works are the great *sine qua non* upon which modern urban civilization has risen.

Part II

EXPLORING MILWAUKEE COUNTY

MILWAUKEE: JUNEAU TOWN (EAST TOWN) AND LAKE PARK

5.1 US Bank Center

777 E. Wisconsin Avenue
Completed in 1973
Architectural firms: Skidmore, Owings & Merrill with Fitzhugh Scott
    Architects; Fazlur Khan, engineer
Foundation: Steel H-piles to bedrock
Geologic features: Tivoli Travertine; Coated Aluminum

Milwaukee's tallest building is the perfect place to begin our exploration of Juneau Town's architectural landmarks. At forty-two stories and perched at the eastern end of Wisconsin Avenue, it serves as an immense Mile Zero marker for that amazing boulevard of countless geologic wonders.

In classic International Style fashion, this building is mostly an essay of metal and glass, but down at ground level, and most visible on the eastern exterior along Cass Street, is a magnificent exposure of 3,500 tons' worth of Tivoli Travertine, a stone of ancient architectural pedigree adorning this paragon of modernism. Quarried in central Italy's Acque Albule region near its namesake town, it is in fact the type of limestone that forms from the accumulation of calcite deposited by calcium-rich water issuing from thermal springs. A favorite of Imperial builders two millennia ago, it graced many famous structures in and about Rome, including the Colosseum, where it's estimated that more than 100,000 cubic meters of this rock type originally covered the exterior. And it's still produced and widely used worldwide today. But while much of the other building stone

MAP 5.1. Sites in Milwaukee's Juneau Town and Lake Park neighborhoods.

types seen in Juneau Town originated millions or even billions of years ago, the Tivoli formed in our present period, the Quaternary. So while in terms of human use it is indeed venerable, it's remarkably young by the grand inhuman standards of geologic time.

What it lacks in striking coloration the Tivoli delivers in its combination of good durability and fascinating texture. Make sure you take a close look at the clusters of large and small *vugs* (holes or pits) loosely arranged in undulating bands that are highlighted by dust that settled on the still-fresh deposits. The vugs themselves were produced by the carbon dioxide bubbles also vented by the springs, with faster flowing water producing a coarser fabric. Here the cladding panels have been carefully set to orient these bands horizontally, just as they were in outcrop. Nowadays, when the Tivoli Travertine is used for flooring or counter-tops, the vugs are filled with epoxy or some other material, but thank goodness that hasn't been done here, where we can admire the rock in its original state of unblemished holeyness.

Above the stone base soars the latticework of window glass and white cladding. The latter, coated aluminum, is in its own way just as much a geologically derived material as the travertine. The most abundant metallic element in the Earth's crust, aluminum is usually extracted from bauxite, an unusual ore rock that often resembles fruitcake even more petrified than the kind I remember from the Christmases of my childhood. It is thought to have formed in many cases from highly weathered soils in hot and humid climates. Another famous

Khan-engineered skyscraper, Chicago's John Hancock Center, is also clad in aluminum, albeit in a much more sober black, anodized form.

Supporting all of this great structure in downtown Milwaukee's squashy substrate is a complex of 500 H-piles, so named because these wide-flanged steel columns are H-shaped in cross section. This design can bear considerably more weight per pile than earlier types could. In this case, they were fabricated to withstand up to 640 tons per pile, whereas a maximum of 100 tons per pile had been the previous permitted limit. Reaching all the way to bedrock, they provide an added measure of stability that Milwaukee's nineteenth-century engineers could only have dreamed of in a time when both the natural length limitations of timber poles and a less powerful pile-driving technology meant stopping at a hardpan layer well short of the uppermost strata of the Silurian dolostone.

5.2   Northwestern Mutual Life Insurance Building

720 E. Wisconsin Avenue
Completed in 1914
Architectural firm: Marshall & Fox
Foundation: Timber piles to a clay layer
Geologic features: Woodbury Granite, Carrara Marble, Bronze, Terra-Cotta

This dignified seat of corporate gravitas, with its set of ten pilaster-flanked Composite-order columns, is not Milwaukee's most grandiose statement of neoclassicism. But it is its most effective. And it's also a geological treasure trove from top to bottom. It is best to start at the latter level, because this site's foundation and substrate have been the subjects of much speculative tale-spinning in reportage and blog posts. All this creative writing has lofted the building's foundation and soggy setting into the stratospheric strata of the legendary.

A survey of these accounts will net you quite an inventory of fascinating and entertaining tidbits. You'll learn that the structure rests, precariously one would think, on a concrete slab perched atop a still-extant body of water named Lake Emily, which these days does not worry itself with a lake's usual need to have a visible surface to avoid being merely an example of groundwater. Alternatively, you'll be told that the building rests on myriads of piles that have been driven or even drilled deep into bedrock. And you will discover that Lake Emily is a subterranean marvel that may just possibly have a ghostly or supernatural aspect; at any rate, it lurks beneath, as indestructible as it is elusive, a spectacular natural wonder that adds to the Cream City's distinctiveness. You may even chance upon a 1952 newspaper piece that quotes an authority, then a curator of the

Milwaukee County Historical Society, who claimed that Lake Emily ultimately proved unfillable because it's fed by an underground river. Clearly, all these little bonbons of urban folklore need to be examined for real nutritional content; to use the oft-repeated Carl Sagan dictum, "extraordinary claims require extraordinary evidence."

To begin with, was there *ever* a Lake Emily? Indeed there was. A historical retrospective published in the *Milwaukee Sentinel* in 1898 describes both its origin and demise in or about the 1840s, some seven decades before Northwestern Mutual was erected. Really much more of a modest pond than a true lake, a portion of it was situated where this building now stands. According to the newspaper account, it came into being when the bluff originally fronting Lake Michigan was partly cut down to provide fill to raise the grade of Wisconsin Avenue and its environs. Evidently this small body of standing water then formed because no culvert was included in the landfill project to compensate for its derangement of the preexisting drainage pattern. For a while little Lake Emily was used recreationally by the local kids, but ultimately, to quote the article, "the city fathers voted the lake a public nuisance, and the lighthouse bluff served to obliterate it." In other words, it was filled in with what was left of the high ground just to the east. Houses were soon built where once children had fished, swum, and sailed their model boats. This is not to suggest that the site of the former Lake Emily was any less saturated with groundwater than the surrounding land; the water table continued to be high this close to the real lake, and this made the underlying sediments here as challenging to heavy construction as anywhere else in Juneau Town. Or even more so. The *Sentinel* even noted that one of these post-Emily structures, a substantial brick double house located just west of the Northwestern Mutual site, was still extant at time of writing and leaned noticeably northward because it had not been stabilized with a pile foundation.

So in any sense does Lake Emily *still* exist? I suppose anyone could dig a hole in the ground, fill it with water, then dump all its soil back in and claim that the hole still exists as an invisible entity imbued with mythic significance. But to do so is to stretch the agreed definition of a hole—or in this case a lake—to the breaking point. And regarding the issue of an underground river, I can vouch that this is, for some arcane reason, a favorite trope generated deep in the Midwestern psyche. For instance, I've now lost count of the number of times residents of Chicago's far-western suburbs, blissfully unburdened by even a shred of evidence, have told me that their well water is especially pure because it has traveled all the way from Lake Superior via a subterranean, trans-Wisconsin waterway. As mentioned above, the Northwestern Mutual site has plenty of groundwater, but any record of it being gathered into just one subsurface channel seems to have eluded the scientific literature. Still, underground-river aficionados can find their dreams come true in the beautiful but geologically dissimilar karst landscape of southern Indiana, where the limestone bedrock is honeycombed with caverns, sinkholes, and other solution features that sometimes result in surface streams detouring underground.

Fortunately, another reliable account of this site, published in the March 1, 1913, issue of the *Engineering Record*, gives us a detailed view of what the contractors actually found when they began to lay the Northwestern Mutual foundation. While there's no mention of their being bushwhacked by an underground lake or river—the name Emily is conspicuously absent—it notes that "when operations commenced in November, 1911, a large portion of the site was covered by the cellars, from 6 to 10 ft. deep, of former buildings, below which the soil consisted of sand and muck, down to ground water level, about 25 ft. below street grade." To keep the excavation dry, *sheet piling* fabricated by United States Steel was inserted around the periphery to form a cofferdam that kept the groundwater out. (Unlike foundation piles, sheet piles are designed not to bear heavy loads but to fit directly together to form an impervious wall.) Then a series of trenches were dug, and, as the article notes, timber "foundation piles, 40 ft. long, were driven 2 ft. apart in staggered rows in the trenches by four steam hammers. . . . The piles were driven to a penetration of 35 to 40 ft., in the sand and blue clay underlying the mud and were sawed off 6 in. above the bottom of the trenches." After that, the trenches were filled with the concrete that would serve as the system of footings for the building's aboveground support of 116 steel columns.

Still, one aspect of the Northwestern Mutual foundation legend most frequently mentioned in popular accounts is perfectly correct: over the years great care has been taken to keep the wooden piles fully submerged in groundwater to prevent disastrous rot that could undermine the structure. This is accomplished by the due diligence of the Northwestern Mutual maintenance staff, and by a special watering system. (For more on that, I heartily recommend Bobby Tanzilo's *Hidden History of Milwaukee*—see the Selected Bibliography.) Regrettably such foresight and attentiveness have not always been manifest at other older buildings downtown, and the sad results are noted in chapter 4. However, the mention in various popular accounts of the piles reaching all the way to bedrock is simply not supported by the *Engineering Record* report. And the 1928 well-construction records for the Northwestern Mutual site and contiguous lots, logged by Knaack & Son Company and still on file with the Wisconsin Department of Natural Resources, demonstrate that only various layers of unconsolidated sediments are present all the way down to 100 or even 160 feet. This of course means that the bedrock lies even farther down, well out of reach of the 40-foot Northwestern Mutual piles. This great depth to bedrock is also typical of other logged Juneau Town locations. Anchoring piles in sand and clay rather than stone was the standard nineteenth- and early twentieth-century practice, and to this day is no cause in itself for concern about their buildings' stability. Piles reaching the Silurian dolostone are characteristic instead of more recent and more technologically advanced projects, from the 1973 US Bank Center and the Couture, very recently completed.

While the oozy subsurface of the Northwestern Mutual building may seem the site's main geological story, there's plenty of interest aboveground as well. Besides the bronze entrance doors and lampposts (good examples, in both weathered and unweathered form, of the metal alloy discussed in the following section), there's the stone-clad exterior. It is this region's best exposure of northern Vermont's Woodbury Granite, a handsome, gray, salt-and-pepper rock of igneous origin that ranges in composition from true granite to granodiorite. Medium-grained in texture, it has black flecks that are mostly the mica mineral biotite, and distinctive white crystals of oligoclase, a plagioclase feldspar. Your close scrutiny

FIGURE 5.1. Milwaukee's finest expression of the neoclassical style, the Northwestern Mutual building is also a showplace for Vermont's Woodbury Granite. And it's the site of both an impressive pile foundation and the unsinkable Lake Emily urban myth. In the background at right rises the Northwestern Mutual Tower, anchored instead on a shallow concrete pad.

will also reveal glassy gray quartz and alkali feldspar that, while in other instances is usually pink or white, in this case can be either clear or very faintly blue. The Woodbury dates to the Devonian period and owes its existence to the collision of Laurentia (ancestral North America) with the microcontinent Avalonia and other wandering landmasses that added to the region that is now New England. As the leading portion of Avalonia's plate subducted under the Laurentian margin, the intense heat, pressure, and chemical differentiation that resulted created gigantic blobs of relatively low-density magma that rose under the newly formed Acadian Mountains. Some of this magma reached the surface to generate volcanic eruptions, but the portion that would ultimately become the Woodbury Granite solidified while still deep underground. Only many millions of years later, long after the disappearance of the Acadian chain, did the forces of erosion expose this *pluton*, or large mass of intrusive rock, at the surface.

The Woodbury Granite, still quarried today, was a favorite of Beaux Arts architects. Both here and on the exterior of Chicago's City Hall and County Building, it harmonizes perfectly with all the Hellenistic grandeur. A marvelous and enduring carving stone, it holds crisp sculpted detail seemingly forever but also imparts the required sense of massiveness, as seen here especially in the façade's 74-foot-tall columns, each of which weighs 422 tons. Behind them, the windowed wall is clad in yet another geologic material, green terra-cotta that nicely complements the Woodbury's lighter cast.

Buildings as lavishly designed and decorated as the Northwestern Mutual are bound to have petrologically interesting interiors, too, though often the exact identities of the fancy stone within are not well recorded for posterity. Luckily, in this case the main selection, adorning the foyer and main-lobby walls, does have good provenance. It's the ne plus ultra of true marbles, the Carrara, quarried from ancient Roman times to our own in the Apuan Alps of northern Italy. Here it is on display in both pure white and white-and-gray-veined forms. What makes this stone real marble, and not just one of the other polishable rock types architects call "marble," is its origin: it's a limestone that was subsequently metamorphosed by the forces of plate compression and mountain building into its present recrystallized form of finely textured calcite grains. When speaking of a metamorphic type like this, it's best to cite not its age but its *ages*. First, its *protolith* (original rock) was a limestone formed on a carbonate platform in an arm of the Tethys Ocean from Late Triassic to Early Jurassic time. Then its transformation into marble occurred when it was buried deep under *nappes* (thrust sheets) of other rock. That episode of crustal compression took place in the Late Oligocene and Miocene epochs.

The Carrara is also present as trim on the lobby floor, though most of that surface is decked out in a nicely contrasting, rose to light-brown selection that closely resembles Missouri's Warsaw Limestone, better known to architects as "Carthage Marble." The Warsaw, which dates to the Mississippian subperiod, has often been used for flooring, so there's a good possibility that this is what it actually is. It's complemented with a striking green stone that is either a serpentinite or its brecciated derivative, ophicalcite. Most likely it was quarried in the Italian or French Alps, or in Vermont. But sadly neither of these rock types has been documented for this site, as far as I can determine.

5.3 Northwestern Mutual Tower and Commons

805 E. Mason Street
Completed in 2017
Architectural firm: Pickard Chilton
Foundation: A hybrid system discussed below
Geologic features: Buddusò Granite, Yule Marble, Botticino Lime-
    stone, Eramosa Dolostone

If the 1914 Northwestern Mutual Building next door is Milwaukee's paragon of neoclassicism, this soaring and gracefully curved mass of glass is one of the city's boldest statements of the early twenty-first century. It's also undoubtedly one of the most fascinating geologic sites in the downtown area—and an engineering marvel as well.

The Tower portion of this connected complex is held securely in place by an old method given new life by modern technology and clever, cost-saving engineering. Whereas most of its tall neighbors young and old rely on foundations of either the pile or caisson type, this thirty-two-story high-rise sits on a base whose thickness and careful construction make up for its lack of deep roots. This is not the first time this method, known as a raft or mat foundation, has been used in recent times; it was also employed for the Quadracci Pavilion of the Milwaukee Art Museum (site 5.40). But that was for a building, for all of its beauty and sophisticated engineering, that is considerably lower and lighter. The Tower's massive pad contains a lattice of 1,350 tons of steel rebar encased in 9,200 tons of concrete that was delivered to the construction site in one continuous, twenty-nine-hour-long pour. That crucial phase involved the meticulously choreographed arrival and departure of over 900 ready-mix truckloads. When blended together, the concrete's components—water, sand, aggregate, and lime-containing Portland cement—trigger a chemical reaction that, being exothermic, releases a great deal of heat. That's an essential aspect of the hardening process, but it's also a potential liability, in that it can cause disastrous expansion cracking.

To prevent this, tubing that circulated coolant water was positioned at the base of the pad, and the temperature was carefully monitored by a network of sensors throughout the weeklong curing period.

But the Commons section is a different, more complicated story. Its western portion benefits from being positioned on the site where two previous Northwestern Mutual additions had successively stood. The original basement and supporting wooden piles, driven in 1931, had been judiciously retained throughout the comings and goings of these structures, and it's this foundation, almost a century old now but still doing its duty, that forms the base of this part of the new four-story structure. And what had originally served as the groundwater drainage network here has been converted into a nifty hydration system that keeps the aging piles completely bathed in groundwater. In that anoxic environment they're much more resistant to rotting. In contrast, the eastern portion of the Commons, which stands beyond the original buildings' footprint, relies instead on shallow spread footings. The Tower's concrete mat is connected to the Commons' hybrid foundation by expansion joints designed to prevent movement between them.

To explore the ornamental-stone types on display at the Tower and Commons, the best place to start is outside, along Mason Street. There you'll find, in addition to all the glass and metal, three-centimeter-thick cladding panels made of a very pale-gray, medium-grained igneous rock. This is the Buddusò Granite, often known to architects and contractors instead as "Sardinian White." Dating to the late Carboniferous or early Permian period, it has been delimited to the span 310–280 Ma ago, which indicates that it began its existence as magma rising to form the great Corsican-Sardinian Batholith now exposed on those two large Mediterranean islands west of mainland Italy. This huge mass of intrusive crystalline rock came into being during the Variscan Orogeny, a mountain-building event triggered by converging crustal plates during the assembly of the supercontinent Pangaea.

Besides being one of the granitoid rock types that gives the modern Sardinian landscape such a distinctively craggy and sometimes otherworldly look, the Buddusò has become in recent decades a darling of American architects: it can be found not only here but on some of Chicago's most notable postmodernist buildings. While it's perfectly accurate just to call it a true granite and leave it at that, igneous petrologists parse it further by calling it either a *leucogranite* (a notably whitish or pale-gray granite) or a *monzogranite* (a granite with roughly equal amounts of alkali and plagioclase feldspars). Here, these minerals are the white and olive crystals; they're complemented with gray quartz and black biotite mica.

Once inside the Commons, you'll immediately be struck by the profligate use of the gleaming white cladding that dominates the scene. This public space is nothing less than the most dazzling and extensive use of true marble in modern architecture I've seen. And that marble is the "Calacatta C" variety of the Yule, every bit as pure and premium-grade as Italy's famous Carrara, for which it could easily be mistaken, even by an expert. Quarried at great effort and cost—its extraction site lies 9,500 feet above sea level, in the heart of a mountain, in one of the remotest parts of the Colorado Rockies—the Yule is a fine-grained and nonfoliated metamorphic rock of exceptional luster and beauty. Its protolith, the Mississippian-age Leadville Limestone, can still be found elsewhere in its unaltered state. But at what is now the quarry locale, its beds happened to be altered by a nearby intrusion of magma. In the Late Eocene and Oligocene epochs intense heat from that body of molten rock essentially cooked the drably bluish-gray Leadville into the glorious, crystalline, snow-white Yule. So highly regarded is this marble that the company now extracting it in Colorado is Italian. Among its on-site staff are representatives of age-old Carrarese quarrier families. Remarkably, given the fact Carrara itself is still producing great quantities of stone, the Yule is mostly shipped to Italian markets. That is a sure sign of its high regard among the world's most discerning and experienced marble producers. But the Yule Marble you see at the Tower and Commons has had an even greater journey. Once removed from its mountain fastness, it was shipped more than a quarter of the way around the Earth to Pietrasanta, Italy, a major stone-processing center just downslope and a bit down the coast from Carrara. Then it was shipped back across the Atlantic for installation here.

Though the Yule Marble is definitely the show-stealer, there are two other notable stone varieties as well. One of these, a popular selection for the interior of grander American buildings for well over a century, is the Botticino Limestone; the other, a much more recent introduction and still a rarity in this region, is the Eramosa Dolostone. The Botticino, present here in its "Semi Classico" brand, is used for flooring pavers. A beige stone quarried since ancient Roman times in northern Italy's Lombardy region, it presents plenty of interest to the visitor with an eye for detail, and here and there contains invertebrate fossils that give us a glimpse of Lower Jurassic marine life in the great Tethys Ocean 200 Ma ago. And it's often marked with wavy slate-gray lines called *stylolites*. These probably formed when layers of lime mud were compressed under the weight of newer deposits. This for some reason caused films of soluble ions within them to migrate from their original positions, leaving behind zones of insoluble substances visibly darker than the surrounding sediments.

Architects typically call the Botticino a marble because it's relatively soft and takes a high-grade polish. But geologists define it instead as an unmetamorphosed chemical sedimentary rock. One indicator that this is true lies in the fact it contains intact fossils, as mentioned above. They're a classic sedimentary feature that only very infrequently survives the recrystallization caused by the high temperatures and increased pressure that true marbles undergo. That's why you can vainly search forever for remnants of ancient life in the Yule, though its own parent limestone has a plentiful supply of them.

The Eramosa Dolostone, though not as extensively used here, is found as accents in the Botticino flooring and also at the base of wood wall paneling found in the Commons. This, too, is a chemical sedimentary rock, but it provides an attractive contrast in both tone and texture, being a deep gray with gently undulating laminations. Quarried on the Bruce Peninsula of Ontario, about 150 miles northwest of Toronto, this rock is Upper Silurian in age, which makes it twice as old as the Botticino. Officially described by Canadian stratigraphers as the Eramosa Member of the Guelph Formation, it's of special interest because it's *bituminous*—saturated with organic matter. As a result, the Eramosa has been studied by economic geologists for its petroleum potential, though to my knowledge it has never actually been exploited as a source of fossil fuels, an energy source that in any event we should have had the sense to turn away from long before this. But its high organic content does indicate that it probably formed in still, shallow, and oxygen-poor saltwater, though other environments are also suggested by various sections of this rock unit. Of approximately the same age as Milwaukee's much lighter-toned Silurian dolostone bedrock, the Eramosa formed on the opposite side of the reef-ringed Michigan Basin and is well known to paleontologists as a *lagerstätte*, a sedimentary rock unit of exceptional importance to the study of ancient life because it contains marvelously preserved fossils of soft-bodied creatures and other organisms not generally found elsewhere. While you're unlikely to find such exotic life-forms on display here, it's still worth recognizing the Eramosa's importance to evolutionary biologists as well as builders.

5.4   Wisconsin Gas Building

626 E. Wisconsin Avenue
Original name: Wisconsin Gas Light Company Building
Completed in 1930
Architectural firm: Eschweiler & Eschweiler
Geologic features: Morton Gneiss, Oneota Dolostone, Bronze

In the first book in this series, *Chicago in Stone and Clay*, I introduced the concept of a "Grand Art Deco Formula." That's my term for the pattern one sees again and again in the design of American skyscrapers built in the 1920s and 1930s, during the flowering of the Art Deco and Art Moderne styles. These elegantly streamlined and set-back towers are in effect mountains of Salem Limestone, a rock type from southern Indiana noted for its dependability, workability, and relentlessly uniform appearance. But the most intriguing aspect of the Formula is that its purveyors usually offset all this blandness of tone and texture with a distinctly contrasting and even subversive element: highly patterned, attention-grabbing igneous or metamorphic rock sheathing the lowest and most visible levels of their buildings. You'll find this exact pattern of stone use on full display farther west on Wisconsin Avenue, at the Warner Building (site 6.7) and the Wisconsin Tower (site 6.11). Here at Wisconsin Gas, famed Milwaukee architect Alfred Eschweiler did not completely eschew the Grand Art Deco Formula, but he did give it quite a twist. And we can be grateful that he did. The result is an unmitigated architectural masterpiece, a perfect harmony of form and color and detail. Milwaukee contains so many notable designs, but is there another that is as immediately pleasing to the eye?

While it's best known for what's at its summit—the flame-shaped lamp that faithfully signals changes in the weather—it's what's at the bottom of Wisconsin Gas that is really worthiest of the urban naturalist's awe. For the exterior of the two bottom floors is clad in a stone known to builders as "Rainbow Granite" and to geologists as the Morton Gneiss (pronounced "nice"). This flamboyant stone, quarried in the amazingly complex geologic setting of south-central Minnesota, is a frequent ingredient in the Grand Art Deco Formula, as indeed you'll see at the other Wisconsin Avenue sites mentioned in the previous paragraph.

The fact that the Morton Gneiss was so often used in that era speaks well of its users and their design sense. Each time I stand here and look at this remarkable rock type, I run the danger of getting permanently lost in its riotous patterns of swirling pink and gray and white and black. Sometimes I seem to be peering into flames at the maw of hell. Or I'm swept along inside a prairie tornado. Clouds and faces and ocean waves take form, oscillate, evaporate, and then are kaleido-scopically replaced by other shapes. No other rock type is half so wild, inchoate, anarchic, hallucinatory.

And no other rock type you'll see in this or any other American city is so ach-ingly ancient. If you stretch out your hand and touch its polished surface, you'll make a fleetingly ephemeral creature's best approach to eternity. In recent years a highly sophisticated technique using an ion probe to date the rock's zircon crys-tals has determined that the Morton's oldest constituents date to about 3.52 Ga, which makes this the only building stone cited in this book that traces its origin to the remote Paleoarchean era. This staggering age deserves both our reverence and our understanding, though in truth no human mind can really comprehend the full implication of this immensity of transpired time. But it helps to consider that this rock is three-quarters the age of the Earth and our solar system, and

FIGURE 5.2. A magnificent twist on the Grand Art Deco Formula, the Wisconsin Gas Building has upper reaches not of the usual buff-colored Salem Limestone but of deeper-toned facing brick trimmed with Oneota Dolostone. That said, its base is still characteristically clad in richly patterned crystalline rock. It offers the urban geologist one of the city's finest exposures of Paleoarchean-era Morton Gneiss.

more than a quarter the age of the entire universe. When the stone first formed, the world was a radically different place. Days lasted only about twelve hours because our young planet spun twice as fast as now. No plants or animals existed or would exist for another three billion years or so, and the Earth's living cargo consisted only of single-celled microbes. The sky above, devoid of the free oxygen we so dearly need, was dominated by a massive Moon that orbited the Earth at less than half its current distance and generated colossal ocean tides. Ironically, though, the Sun was notably fainter, being only about three-quarters as luminous as it currently is.

In the long span since, the Morton Gneiss has been subjected to an unimaginable amount of geologic Sturm und Drang, and has even come close to being wholly returned to the state of incandescent magma. Its complicated and violent history is reflected in the fact that it really isn't just one type of gneiss, or any other single type of rock. It's what petrologists call a *migmatite*—a "mixed" or hybrid rock, a collage of different rocks, really, that bears the scars of partial remelting and the creation and intrusion of younger types within it. In the Morton these different varieties include gneisses derived from tonalite, granodiorite, and other felsic igneous rocks; black amphibolite, which is metamorphosed komatiite or similar mafic rock; and crosscutting dikes of fine-grained and igneous aplite that, at the mere age of 2.59 Ga, are the newest pieces in this petrologic puzzle. However, if we look not at its list of ingredients but at its larger context instead, we'll see the Morton belongs to a fragment of early and primitive crust, the Minnesota River Valley Terrane, that significantly predates the formation of the Superior Craton, with which it ultimately collided to form what is now the latter's southwestern margin, at about 2.68 Ga.

Here on the Wisconsin Gas exterior the Morton demonstrates its ability to be sliced into large cladding panels, polished to a high sheen, and carved into gracefully curved surfaces. Make sure you also take a good look at the large clasts of jet-black amphibolite that float here and there like islands of night in the lighter-toned gneisses; they're a signature feature of this famous stone.

Unfortunately, the identity of the darker igneous rock that serves as the base course and door-frame trim was apparently not recorded. (It reminds me of the "Charcoal Black" version of the St. Cloud Area Granite, also from Minnesota, but without provenance it's impossible to be sure.) Still, no such ambiguity of identity exists for the magnificent sunburst that adorns the entrance. It's made of that beautiful and venerable alloy, bronze, which has been in continuous human use for at least seven millennia. In its own way every bit as geologically significant as the stone and brick it complements, it's derived from rocky ores of copper and tin, and sometimes of antimony and other metallic elements as well. What you see here is in fact a 2003 replica of the original ornament, which had been removed in 1966. Still a gleaming golden yellow, it has been treated to prevent weathering to a duller green or brown patina.

If the two lowest floors of the exterior showcase materials that fit well within the Grand Art Deco Formula, the fact remains that the handsome mass of brick rising above them is a notable departure from it. The geologist may regret that the source of the brick was not recorded, but no matter. How well the mottled pattern of ocher and cream works in imparting greater visual detail and a warmer tone than the standard Salem Limestone could. And its impact on the beholder is heightened by the yellowish-tan stone that's used for ornament and trim in the upper reaches of the setbacks. This is a sedimentary rock type, the Oneota Dolostone. While later on we'll see it much closer at hand on other notable Milwaukee buildings, we must be content here to admire it from afar. Much better known in the building trades as "Kasota Stone," this distinctively tinted rock type is Lower Ordovician in age, which means at a little less than half a billion years old it's about one-seventh the age of the Morton Gneiss. But like the latter, it hails from Minnesota. While the Oneota has been quarried in several locales in the southeastern portion of that state, the stone here comes specifically from the town of Mankato, 65 miles southwest of Minneapolis.

5.5 Northern Trust Building

526 E. Wisconsin Avenue
Original name: Northwestern National Insurance Company Building
Completed in 1906
Architectural firm: Ferry & Clas
Geologic features: Salem Limestone, Bronze, Terra-Cotta Roof Tile

Our second example of the Beaux Arts style along East Wisconsin Avenue is a decidedly smaller building than the current Northwestern Mutual headquarters two blocks to the east. But it too is a first-class showpiece of the neoclassical, which at the onset of the twentieth century was the look preferred by so many banks and insurance companies eager to portray their institutions as rock-solid temples of commerce and financial stability. This was also the time, in the wake of the Spanish-American War, when references to the grandeur of ancient Rome suggested, subconsciously at least, a similar imperial destiny for this nation.

All that noted, the Northern Trust is also a prime exemplar of the one rock type that has since the late nineteenth century come closest to dominating the American building-stone market. This is the Salem Limestone, better known to builders, historians, and its own producers as "Bedford Stone" or "Indiana

Limestone." Any geologist who has explored Milwaukee or practically any other city, suburb, or country town in this country soon develops something of a love-hate relationship with the Salem for the simple reason that it's everywhere. At times it seems to be the lithic equivalent of the houseguest who doesn't know when to go home, or the weed species that outcompetes the native flora almost to the point of extinction. But this ubiquitous Hoosier stone also gives us a chance to see how one type of rock performs used in a plenitude of places, styles, applications, and microclimates. In Milwaukee, for instance, it constitutes the primary exterior material of a host of major buildings; it performs a supporting role as trim and carved ornament for old churches of locally quarried stone; and it's even found in cemeteries as intricately sculpted, allegory-laden monuments to the dearly departed.

Apparently the Salem was first quarried in Stinesville, Indiana, in 1827, but by the following decade it was also being extracted farther south, in such Lawrence County communities as Bedford, a town destined to be famous as the center of its production. Still, for a long time thereafter its use was mostly local. Things finally began to change in or about 1871, when Chicagoan John Rawle visited the area. So impressed was he with the Salem that he returned home to open his own stone yard and market it to what turned out to be an eager clientele. For his timing was perfect: at that juncture Windy City architects were desperate to get their hands on any and every suitable building-stone type they could find. After all, this was the frantic rebuilding boom following the Great Fire. Soon the Salem was eclipsing even the native Silurian dolostone in popularity. For one thing, it was found in massive beds that yielded ashlar, panels, and column shafts of any desired size or shape. And, for another, it was highly appreciated for being what masons call a *freestone*—a rock type that unlike most others can be successfully sawed or split in any direction without unwanted fracturing. As geologist George Perkins Merrill noted in 1891, it had the ideal properties of being "soft, but tenacious." A sculptor's dream, it seems to have been expressly made for the chisel; it graciously takes intricately carved detail that often remains crisp and pristine for decades thereafter.

And from a purely geological standpoint, the Salem is notable for a number of other things. It's one of the lower Midwest's more widely distributed sedimentary formations, and can be found outcropping in western Illinois and Kentucky as well. It dates to the Mississippian subperiod of the Carboniferous, a particularly interesting time in the history of what is now North America's midsection, when much of it was covered by a shallow *epeiric* (continent-covering) sea of warm saltwater. At that time our region lay close to the equator, and in this tropical climate the sea was home to a vast population of calcite-secreting organisms that ultimately produced huge quantities of limestone. The Salem in particular formed on a carbonate platform in shoals, tidal channels, and lagoons teeming with life. A *biocalcarenite*, it's composed of small grains of organic debris—a multitude of broken shell fragments, some small fossils still intact or mostly so, and the microscopic tests of the single-celled foraminifer *Globoendothyra baileyi*. While some other Mississippian limestones are *packstones*—they have a greater mud content and are more compact and smoother in texture—the Salem is classified as a *grainstone*. In fact, its grainy texture is quite similar to the feel of a sandstone.

As the Salem became ever more popular and widespread in the late 1800s, it developed a record of endurance and dependability. While it can't be polished the way some other limestones can, and while it isn't immune to the accumulation of soot, it has the remarkable quality of often weathering very little, or not at all. Its base color of pale gray to buff simply doesn't alter over the passing of the decades. If this wonder stone does have an Achilles' heel, though, it's its porosity. When it comes into contact with the ground, it wicks up water and dissolved wintertime deicer salts, which then recrystallize between its grains and cause bad spalling and cracking as well as an unsightly white surface crust known as *efflorescence*. Clever architects obviate this common problem by providing a base course of less permeable igneous rock.

And there's also the matter of this limestone's overall visual impact, or rather nonimpact. In direct contrast to the maniacally patterned Morton Gneiss on display one block down at the Wisconsin Gas Building, the Salem is so uniform and unassuming in aspect that it borders on invisibility. I often ask my urban-geology tour participants if the Salem exposure in front of them is in fact stone or just a fine concrete. Most guess it's the latter. Still, in architecture there are times when a material's understated nature is its major asset. And on a planet as reactive and corrosive and prone to change as ours, all rock types, and indeed all materials exposed to the elements, have their vulnerabilities. In the end Mother Nature, patient and relentless, degrades and consumes them all. That said, what's so admirable and remarkable about the Salem is that its popularity and glowing reputation as the workhorse of American architecture has never really waned since the rise of the Beaux Arts esthetic and its preference for lighter-toned stone. It has effortlessly survived so many changes in taste and style.

If you take the time to really scrutinize the Salem on the Northern Trust façade, you'll spot interesting details that are some of its most salient identification traits. The main ashlar blocks have smooth-sawn faces where, with the help of a hand lens or pocket magnifier, you can better recognize and appreciate the stone's fabric of tiny shell fragments. And just to the east of the main entrance you'll find one surface that has for some reason weathered more, to reveal patterns of slightly curving horizontal lines. This is *crossbedding*, a feature formed by sediments deposited in such high-energy environments as windy sand-dune fields and flowing streams. In the Salem's case, however, the pattern points to grains agitated by near-shore surf or currents in a tidal sluiceway. Also take a moment to look above, at the bases of the fluted Ionic columns, where the intricate carved detail is still as fresh as the day it was made. It's a terrific example of the resilience and ornamental potential of this stone.

Having already noted the brightly metallic, unpatinated form of bronze in the preceding section, I'd be remiss not to pay homage to one of its weathered forms here, on display in the elegant lamppost and the door and window grilles. The alloy has been allowed to freely react with its environment, and the result is a dramatic color change that signals the presence of either copper-carbonate or copper-sulfate compounds. These produce the striking green hue that imparts both a self-sealing coating to the bronze and an air of dignity and venerableness to the design. On the other hand, it also nicely spices up the subtle earth tone of the Salem. The same can be said for the bright brick-red roof tiles of the third, dormered story. To see these, you'll need to cross Wisconsin Avenue. While the source of these terra-cotta tiles has not been identified, it's safe to say they were fashioned from high-quality clay mined from either glacial till, Illinois Basin coalfield shale, or lake or stream deposits. In the firing process, they were transformed by the searing heat of the kiln from loose sediment into the fabricated equivalent of metamorphic rock.

5.6 Federal Building

517 E. Wisconsin Avenue
Alternative name: United States Courthouse and Federal Building
Original northern section completed in 1899; southern additions featuring other, unidentified stone were completed in the 1930s and 1940s
Architects: Willoughby J. Edbrooke (supervising US Department of Treasury architect); James Knox Taylor
Geologic features: Athelstane Granite, Mount Waldo Granite

Besides being a superb open-air museum of global geology, East Wisconsin Avenue is a veritable cavalcade of America's greatest architectural modes. And here at the Federal Building we come upon the Richardsonian Romanesque, which must by rights be the nearest and dearest style to the rockhound's heart. First formulated by Louisiana-born and Boston-based Henry Hobson Richardson, this esthetic was rooted in the phase of Western European medieval architecture that preceded the Gothic. Sadly, Richardson himself never designed any Cream City buildings; the closest still standing is Chicago's Glessner House, a rock-ribbed masterpiece well worth the trip to see it.

Richardson's mythic vision of buildings rising out of the bedrock below them featured stout semicircular arches, squat columns, rock-faced walls, and brooding towers and turrets. Certainly it was rooted in the Old World past, yet it had and still has a surprising applicability to this younger culture: it strikes a deeply sympathetic chord in the American psyche. Other architects were not slow to recognize this. Though Richardson died in 1886 at the tragically young age of forty-seven, his contemporaries enthusiastically adopted and emulated his ideas till tastes changed near that century's close. Then the Beaux Arts version of classicism swept all before it, and the beautiful was tasked to replace the sublime. Symmetry supplanted asymmetry; rationality and order eclipsed the chthonic rumblings of the subconscious mind. On this one block in Milwaukee you can see the swing in the weathervane of style as you glance from this massive edifice back to the Northern Trust building, described in the previous section and just across the street.

Many Richardsonian Romanesque buildings feature dark-tinted or at least strikingly colored stone. This often adds powerfully to the desired effect, as

FIGURE 5.3. The Federal Building demonstrates how effectively lighter-toned stone can express the Richardsonian Romanesque style. Two gray granites have been used for its exterior. The darker base, including the entrance's column plinths, is Wisconsin's own coarse-grained Athelstane Granite. Above it is the paler, medium-textured Mount Waldo Granite, quarried in Maine. Note the intricately carved ornamental patterns in the latter.

St. Paul's Episcopal Church (site 8.9) certainly demonstrates. But here we see how effective this style can also be when set in decidedly lighter and less attention-getting tones. Still, even from a considerable distance it's obvious that the Federal Building, while clad mostly in a rather neutral light gray, has an exposed basement story that sports a deeper version of that shade. As you approach the façade, you'll also see that the lower stone, while sharing the same pattern of macroscopic crystals typical of granitoid rock, has a considerably coarser texture. This is the Badger State's own Athelstane Granite, taken from the town of the same name a little more than an hour's drive north of Green Bay.

The Athelstane came in three varieties of varying color and grain size; the other two were actually quarried in Amberg, a few miles away. Radiometrically dated to approximately 1.84 Ga, it's Paleoproterozoic in age and hails from a fascinating piece of northern Wisconsin's complicated mosaic of crustal components. This section, known as the Pembine-Wausau Terrane, was an arc of volcanic islands that collided with the edge of the Superior Craton. It was in this context that the Athelstane formed from a cooling mass of magma beneath the once lofty but now utterly vanished Penokean Mountains. The rock type is still formally listed as the Athelstane Quartz Monzonite, though more recent interpretations describe it instead as ranging from granite to granodiorite. (In this book, for the sake of convenience, I just call it granite.) The stone you see on the Federal Building's façade is the grayer of the Athelstane's two coarse-grained varieties; the other, from Amberg's now-defunct Aberdeen and Pike River Quarries, is a much deeper pink.

This granite's big crystals provide an excellent opportunity for some applied mineralogy. The best place to study them is on the dressed-faced plinths of the clustered archway pillars that stand at the main entrance. They comprise roughly equal amounts of the light gray to slightly pinkish alkali feldspar microcline, a white plagioclase feldspar that's probably albite, glassy quartz, and black biotite mica and hornblende.

While we'll see the Athelstane at other Milwaukee sites as well, the stone above it, which constitutes the vast bulk of the building's exterior, is a genuine rarity in these parts. Produced in the town of Frankfort, Maine, near the western bank of the Penobscot River, the Mount Waldo Granite is a finer-grained igneous rock composed of the alkali feldspars orthoclase and microcline and the plagioclase feldspar oligoclase, all more or less white. They are joined by gray quartz and biotite that provides the dark flecking. Though this site may be the Mount Waldo's only showplace in this city, it seems that its quarriers were adept at getting a hefty number of other government contracts as well, because the stone was used in quite a few large and notable public structures. One example was Chicago's own Federal Building, completed in 1905 and since demolished, where no less than

500,000 cubic feet of the granite were installed. It also was chosen to bedeck both New York City's Municipal Building and the Philadelphia Mint.

Like many of Maine's other commercial granites, the Mount Waldo dates to the Devonian and is therefore less than a quarter the age of the Athelstane. The Devonian was an important time in the geologic history of Maine and the rest of New England. In that period the microcontinent Avalonia collided with the margin of Laurentia, the geologic term for that phase of ancestral North America, to produce the Acadian Mountains and the larger landmass of Laurussia. In the process, large bodies of felsic magma rose up from deep in the crust to form plutons. Though the molten rock cooled and solidified into granite before reaching the surface, many of these plutons, including the Mount Waldo's, have since been exposed by erosion and now crop out on or near the Maine seacoast.

5.7   Milwaukee Club

706 N. Jefferson Street
Completed in 1883
Architects: Daniel H. Burnham of the firm of Burnham & Root (principal
    designer); Edward Townsend Mix (who supervised its construction)
Geologic features: Philadelphia Brick, Lake Superior Brownstone,
    Terra-Cotta

In a neighborhood dominated by buildings of yellow, buff, and gray, the Milwaukee Club is, despite its modest size, a ruddy-red Queen Anne standout. Though it was once attributed solely to local architect Mix, it's now acknowledged that it's mainly the design of a world-renowned Chicago architect who, at this phase in his career, left most of the creative work to his brilliant partner John Wellborn Root. But here, apparently, the plans were penned by Burnham's own hand.

There's no question that the most visually impactful part of the exterior is the facing brick, which is ornamented with spandrels and decorative panels of terra-cotta of uncertain source. Though for years I guessed that the brick itself was probably either from Philadelphia or Baltimore—both cities were renowned in the 1800s for the quality of their red brick—I couldn't confirm it anywhere in the architectural literature. But finally the golden day came when I chanced upon a slim volume, a 1982 history of the club written by well-known Milwaukee architectural historian Russell Zimmermann. It confirmed that of my two hypotheses, my made-in-Philly version was correct.

For Midwestern building owners or architects of the nineteenth century, the inclusion of Philadelphia Brick in their designs assured them premium boasting

FIGURE 5.4. A standout in a town proud of its own cream-colored brick, the Milwaukee Club is decked out in its richly red Philadelphia competitor, garnished with Lake Superior Brownstone and terra-cotta.

rights. A top-dollar material, it trumpeted to all that no expenses had been spared in creating an aristocratically handsome and enduring façade. Nevertheless, in a city dominated at the time by its own highly regarded Cream City Brick, the flagrant use of an Eastern competitor, however illustrious, required some serious moxie as well as a lot of extra cash. While it certainly wasn't the only Philadelphia Brick structure in town, the Milwaukee Club must have been the most prominently situated of the lot, and quite a conversation piece when it was completed.

Philadelphia's brickyards benefited from the fact that they lay between two great rivers, the Delaware and the Schuylkill, which earlier in the Quaternary period had deposited thick beds of alluvium known to geologists as the Trenton Gravel. The brick you see here most likely was fabricated from it. The Trenton's formation name is somewhat misleading because in addition to gravel it contains beds of sand and compact, pebble-free clay. The last of these is found closest to the surface, where it was ultimately dug up in vast quantities, especially in the area of the city known as the Neck, where the land between the two south-flowing streams is the narrowest. There the clay's composition was ideal for the manufacture of highest-quality facing brick.

Below the Club's smooth upper façade, and also serving as its trim and sills and lintels higher up, is a maroon stone that is rock-faced in the exposed basement story. In the architectural literature it's identified only as Lake Superior Brownstone. This catchall term indicates it is either Upper Michigan's Jacobsville Sandstone or a similar selection from the Bayfield Group sandstones of northern Wisconsin. The latter shared a similar origin with the Jacobsville, though geologists are still unsure if they directly correspond to it. In any event, the Lake Superior Sandstone you see here, whatever its exact formation name and quarrying locale, came into being at some point after the formation of the Midcontinent Rift, a huge trenchlike structure filled with igneous rock and topped with sedimentary strata that runs through the Lake Superior region. The strange origin and significance of this great crustal feature is more fully discussed in chapter 2.

Lacking fossils and other evidence that could better pin down its age, geologists have so far only been able to constrain the various types of Lake Superior Brownstone to the half-billion-year span from the late Mesoproterozoic to the early Cambrian. While it's a common sight on nineteenth-century Milwaukee Buildings, which are variously adorned with its Jacobsville and Chequamegon Sandstone varieties, it's worth noting that another design of the Milwaukee Club, submitted by the Chicago firm of Silsbee & Kent, featured Connecticut's Jurassic-period Portland Sandstone instead. The Portland, famous as the rock type used in New York City's brownstone neighborhoods, is quite similar in appearance to the Lake Superior stone that is much more prevalent in the Upper Midwest.

5.8   Pfister Hotel

424 E. Wisconsin Avenue
Completed in 1893
Architectural firm: Henry C. Koch & Company
Geologic features: Wauwatosa Dolostone, Salem Limestone, Cream
    City Brick, Terra-Cotta, unidentified serpentinite

In a survey of Milwaukee's surviving inventory of nineteenth-century architecture, a certain pattern of geology and design emerges. There seems to be what I'll simply call the Classic Milwaukee Formula, where native stone and clay are employed in three different but related ways that give this cityscape a look distinct from all others. In the First Version of the Milwaukee Formula, buildings have exteriors made exclusively of Wauwatosa Dolostone; in the Second, they're completely clad in Cream City Brick. But in the magically synthetic Third, there's a combination of the two. And the Pfister Hotel is a premium example of that.

No doubt there are other cities lucky enough to have two locally derived building materials that are both reliable and attractive. But it would be hard to imagine any other town with a coupling that works as synergistically as Milwaukee's does. The comely combination of Wauwatosa Dolostone and Cream City Brick, so instantly pleasant to the eye, is the happiest marriage of rock and clay in all of American architecture. It's a game of subtle contrasts, where the most cheerful shade of yellow rises over a base of pale-gray solidity.

First, let's examine the latter component. A good place to start is at the southeastern, Wisconsin-and-Jefferson corner of the building. Here the Wauwatosa is present in large, rock-faced ashlar blocks. This type of finish may resemble rough, unaltered chunks of stone, but actually it's been skillfully worked with a chisel

FIGURE 5.5. If one can forgive and forget the awful bongo-drum add-on at its rear, the Pfister Hotel offers one of the finest expositions of the Classic Milwaukee Formula. Here Cream City Brick rises above a stolid, rock-faced base of Wauwatosa Dolostone. Also present for duty are Salem Limestone trim, an unidentified serpentinite, and terra-cotta ornamentation.

so that the outer face of each block has relatively recessed margins and a project-ing central section with irregular concavities. The rock-faced look was especially favored by purveyors of the Richardsonian Romanesque; it imparts through the mason's artifice a sense of the wild, rugged, and natural, and almost makes its building appear to be rising out of the underlying bedrock.

Quarried just a few miles west of here, the Wauwatosa Dolostone is of Silurian age and taken from what stratigraphers call the Racine Formation. It formed approximately 425 Ma ago, when this region was covered with a shallow sub-tropical sea, on the reef-studded margin of the Michigan Basin. Racine strata are paleontologically world-famous for the abundance of marine-invertebrate fos-sils they contain, and bits and pieces of these organisms can often be seen in the Wauwatosa. When used in this Third Version style the rock usually just makes up the plinth (exposed basement story), and the brick constitutes almost all of the façade. But at the Pfister the stone has been allotted a more dominant role, and it rises a third of the way up to the roofline. As a result, the hotel presents a par-ticularly substantial, fortresslike aspect.

On its high perch of stone the Cream City Brick isn't available for close inspec-tion, but it does still demonstrate its overall impact, especially when viewed from across the street. And there you'll also notice how effectively it's highlighted by ornamental terra-cotta of a tint slightly more orange. The geologic origin and history of this iconic brick type is more fully discussed in the description of the Keenan Townhouse (site 5.22).

One other item of great interest can be found in the stone panels below the hotel's ground-floor windows, near the Salem Limestone belt course. These are cut from stone architects call "green marble," though it is in fact a different meta-morphic rock type known as serpentinite. Regrettably, it's unclear due to lack of documentation whether this serpentinite comes from Vermont, the French Alps, or northern Italy—these three widely used varieties are so similar in appearance that they're often indistinguishable from one another. But regardless of where it was quarried, the stone you see here has one of the most exotic origin stories in all of geology. It began as *dunite*, a type of ultramafic igneous rock that consti-tutes the uppermost mantle under the thin crust of ocean basins. At some point it was chemically altered when it came into contact with water—either as part of a sinking slab in a subduction zone, or when it was scraped up from the mid-ocean floor—and then pushed onto the margin of a continent in the process of plate convergence and mountain-building. The pressures and temperatures of this last phase altered the rock even more into this lovely deep green criss-crossed with pale veins of calcite or marcasite. Unlike the local materials that stand above it, this stone has taken an epic journey, from mantle to crust and from ocean to land, to get here.

5.9   Conroy Building

725–729 N. Milwaukee Street
Completed in 1881
Architect: Unknown
Geologic features: Unidentified brownstone; unidentified red brick,
    unidentified Terra-Cotta

There are times when the urban geologist, hell-bent on identifying every last stone and brick that stands before him, has to acknowledge an intractable mystery with a lingering sigh of acceptance. But at this site the lack of full understanding is compensated with the unexpected joy of suddenly coming upon this little corker of a Queen Anne confection. It's somehow all the more appealing for being sequestered in a middle lot of a side-street block. On its façade there are so many things going on that one wonders if its designer—whose name is also a mystery—suffered from an advanced case of *horror vacui*, an unbridled fear of unadorned surfaces. If so, the horror has been put to good use.

At first glance the Conroy Building brings back memories of the Milwaukee Club (site 5.7). As it should. There's the same prevailing richness of reds in rock and fired clay. Does the brownstone hail from a Lake Superior quarry? Does the brick come from a Philadelphia clay pit? If I were forced to bet, I'd say yes on both counts. Still, I've seen so many surprises in how and where architects and builders obtain their materials, I'll have to hedge that bet until unequivocal proof presents itself.

5.10   Railway Exchange Building

229–233 E. Wisconsin Avenue
Original name: Herman Building
Completed in 1901
Architectural firm: Jenney & Mundie
Geologic features: Northwestern Terra Cotta, Cast Iron

The Railway Exchange Building may look overly slender, bare-sided, and a bit forlorn among its much lower companions, but that's because its original developer's plan to balance it with a western addition of similar height never materialized. Nonetheless, it still inspires plenty of architectural interest just as it is. For one thing, its chief designer, William LeBaron Jenney (1832–1907), is today famous as the man who designed Chicago's Home Insurance Building, the world's first genuine, steel-framed skyscraper.

Geologically, the Railway Exchange is notable as one of Milwaukee's best showplaces of fired-clay ornament produced by the highly skilled artisans of the Windy City's leading producer of architectural terra-cotta, the great Northwestern works. While local, till-derived clays were the source of the very widely utilized Chicago Common Brick, the Northwestern Company drew much of its base material, shipped in by rail, from the coalfields of the Illinois Basin. There, cyclically repeating strata dating to the Pennsylvanian subperiod yielded, in addition to the soft bituminous coal that fueled the burgeoning cities of the Midwest, beds of both *underclay* and marine shale.

Underclays, situated directly under the coal layers, are in fact *paleosols*—the remains of ancient soils that supported the tropical rainforests of the supercontinent Pangaea some 310 Ma ago. Essentially original sediment that was never transformed into rock, these beds often retain traces of the roots of the giant, spore-bearing swamp trees that grew so rampantly in them. Though somewhat more lithified, the marine shales found elsewhere in the vertical sequence of Pennsylvanian strata are also composed of clay particles that, easily ground out of their crumbly matrix, make an admirable medium for terra-cotta. And, like the paleosols, these shale-derived clays are already free from the pebbles and other unwanted ingredients commonly found in glacially deposited equivalents.

In its heyday, the Northwestern Company produced cladding and highly detailed ornament for the era's most famous architects—Jenney, Louis Sullivan, and Daniel Burnham included. Terra-cotta's advantages over building stone, a material it could closely mimic, included its much lighter weight, its reduced shipping and installation costs, and its ease of cleaning. This last attribute was especially appreciated in the age of coal-burning furnaces and power plants producing a currently unimaginable amount of air-polluting, surface-staining coal soot. And, to top it all off, terra-cotta, born in the infernal heat of the kiln, is a truly fireproof substance, in marked contrast to the stone and cast iron that failed so dramatically in Chicago's 1871 Great Fire. All that said, it does share one thing with all other building materials. In a wet, windy, and caustic world mantled by an oxygen-rich atmosphere, terra-cotta and its metal fasteners eventually degrade, however well mounted and subsequently cared for. So, in common with so many other buildings now over a century old, the Railway Exchange's elegant burnt-clay skin has required significant restoration and replacement in recent years.

Here the somberly attractive maroon Northwestern Terra Cotta and its carefully duplicated modern equivalents are most visible from street level on the lowest three of the building's twelve stories. Above it, facing brick of undocumented origin, also maroon, predominates. And though it's easy to overlook, you'll also find a plinth of painted cast iron, making a rather rare appearance as a skyscraper's basal element. This material, also geologically derived, is discussed more fully in the following site description.

5.11   Iron Block

205 E. Wisconsin Avenue
Alternative name: Excelsior Block
Completed in 1861
Architect: George H. Johnson
Foundation: Shallow; inverted-arch type
Geologic feature: Cast Iron

One of the most glorious aspects of Milwaukee's history is its quality of absence: specifically, the absence of a devastating, widespread event that erased much of its early architecture. Chicago, of course, was not so lucky. Many of its pre-1871 structures and their sometimes remarkably exotic building materials were destroyed in its Great Fire. So nowadays the urban explorer is hard-pressed to find, in that otherwise geologically fascinating place, surviving examples of its first three-and-a-half decades as an incorporated city. For that reason I've told my Windy City students and tour participants that a pilgrimage north of the state line is not a good idea; it's an absolute obligation. Like all great towns, Milwaukee teaches the inquisitive mind much more than its own local history.

And this pilgrimage into the world of the mid-nineteenth century should start here, at the Iron Block. To gaze at its fabulous form from across the street is to slip seamlessly into the muddy, sooty, horse-drawn, pre-electrified years leading up to the American Civil War. Then some of the country's fanciest buildings sported exteriors not of masonry but of prefabricated cast-iron units of intricate design. These had often been forged in far-distant foundries hundreds of miles from the sites of their intended use. Here, the Wisconsin Avenue and Water Street façades were created in New York City's Architectural Iron Works, where George Johnson devised this specific design inspired by the Venetian Renaissance style. Once completed, the sections were shipped west by Great Lakes schooner.

The Iron Block's exterior ornament, replete with grapevines, lion heads, Composite columns, and vermiculated rectangles, certainly constitutes one of the most attractive human uses of the second most common metallic element found in the Earth's crust. (The first most common metal, in case you're wondering, is aluminum.) While in modern times iron is mostly extracted from Precambrian Banded Iron Formation units in such famous mining districts as Minnesota's Mesabi Range, earlier generations of Americans turned instead to "bog iron"

FIGURE 5.6. An emissary from Milwaukee's earlier days, the Iron Block wears a prefab façade that's an impressive lesson in nineteenth-century cast-iron artistry. The building is also notable for its unusual, inverted-arch foundations.

extracted from wetland environments, and also to iron-rich *oolite* deposits—including eastern Wisconsin's own Neda Formation. This unusual sedimentary rock type, situated in Ordovician beds just below the Silurian contact, was once quarried in Dodge County, about 45 miles to the northwest of this city, to supply Milwaukee's nascent ironmaking industry.

Cast iron—that is, iron of relatively high carbon content that needs to be completely melted and then set in molds to form desired shapes—is impressively strong when compressed but quite weak under tension (when pulled apart). It was also frequently used in the 1800s for structural members of major buildings, including the first generation of skyscrapers. Though I've come across one modern architectural history that rather rashly touts its resistance to fire, cast iron not sheathed in protective terra-cotta is by no means invulnerable to the fearsome temperatures generated by major urban blazes. In 1871, for example, many iron-fronted Windy City buildings fared as dismally as those of stone. While it has a melting point of about 2,200 degrees Fahrenheit, cast iron begins to lose structural competence much sooner, at just a little above 800 degrees. Even so, when used in settings such as this, it's so pleasingly evocative of Milwaukee's formative years we can only regret that the Iron Block is this city's sole surviving showpiece of this great phase of American architecture.

Interestingly, the Iron Block is also the Cream City's solitary documented instance of the use of a shallow foundation of the inverted-arch type, which was employed under the building's heavy, ironclad western and northern walls. Here piers of coursed stone rise between and are braced with upside-down circular arches made of brick. Reportedly, this unusual technique was deemed especially useful in supporting weighty structures erected atop soft substrates. Of other still-extant sites with inverted-arch foundations, the Corbin Building, in lower Manhattan, is probably the most notable.

5.12   100 East Building

100 E. Wisconsin Avenue
Original name: Faison Building
Completed in 1989
Foundation: Oil-well casing steel piles; a few surviving Pabst Building
    wooden piles (see discussion below)
Architectural firm: Clark, Tribble, Harris, & Li
Geologic features: Edwards Limestone, Concrete, Alicante Limestone,
    Fumane Breccia

Whatever its merits, the International Style of architecture was rather lithically depauperate. When it came to the exteriors of its buildings, rock and fired clay often occupied a minor role at best. Glass and metal prevailed. But with the rise of postmodernism in the final third of the twentieth century, building stone made a tremendous comeback, with architects employing varieties from around the world, including quite a few that had not been seen in the American building trades before.

Undoubtedly, 100 East is one of Milwaukee's best-known and most highly visible examples of the postmodernist style. On its riverfront site once stood three other notable structures: first, the cabin of Solomon Juneau, one of the city's founders; then the Ludington Building; and then the much taller and heftier Pabst Building, Milwaukee's first skyscraper. It was demolished in 1981. In its arches, gables, and cupola, the current edifice pays more than passing reference to the Pabst Building's design. And in fact a little bit of that old edifice is still here. While many of its hundreds of timber piles were removed to lay the new foundation, some were actually kept in place and just worked around. Each of the new piles consisted of a cluster of a dozen or so recycled oil-well casings—steel tubes that had previously formed protective outer cylinders for petroleum-bearing tubing. These were here driven to bedrock. To keep the Milwaukee River from flooding the excavation, a watertight wall of sheet piles was temporarily constructed at water's edge.

But now what is much more visible than its underpinnings is the building's external cladding of the "Cordova Cream" variety of the Edwards Limestone. It covers the precast-concrete exterior from ground level to 100 feet up. Above that, the concrete itself is exposed. If you look at the tower from a little distance you should be able to see where the stone ends; the color changes from white to buff. While the domestic origin of the Edwards might seem to rule out its status as a selection exotic enough for flamboyant postmodernist tastes, it's the perfect choice here—and a rare enough sight in this city for it to be a genuine geological standout.

Striking from a distance, this limestone becomes even more so when seen at close range, especially if you're paleontologically inclined. Lower Cretaceous in age, it's quarried in the vicinity of Liberty Hill, Texas, some 30 miles northwest of downtown Austin. This type of carbonate rock is what sedimentary petrologists term a *grainstone*—a limestone formed from grains rather than from finer-textured lime mud. Those particles are cemented together with a more coarsely crystalline form of calcite called *sparite*. And the Edwards is also notably fossiliferous, with the remains of such marine invertebrates as echinoderms, brachiopods, and strangely shaped bivalves call *rudists*. Unlike most modern clams and oysters, rudists took on odd curved or columnlike shapes that sometimes resembled coiled snail shells and horn corals.

The Edwards Limestone formed in the warm and shallow saltwater of a carbonate platform during the episode of global high sea level known as the Zuñi Sequence. At this point, the Pangaean supercontinent was already in the process of breaking up, and the earliest rendition of the Gulf of Mexico had formed. And it was the rudists, rather than true corals, that built reefs in this realm. Their reign was relatively brief, however; apparently their entire group disappeared for good during the same Cretaceous-Paleocene extinction event that did in the dinosaurs. Here on the lower cladding panels facing the river and Wisconsin Avenue you should be able to spot pits, vugs, and outlines of fossils at least somewhat suggestive of rudists. But these oddball mollusks can be difficult to distinguish from other organisms and structures found in the rock.

A stop inside to see the lobby is also a high priority, because here we have one of the city's best documented examples of the postmodernist penchant for fancy Old World stone. While the walls and columns are the same Edwards Limestone seen on the façade, the flooring features two Mediterranean Basin selections, both Jurassic in age. The darker, salmon-colored stone, abundantly veined with white calcite, is the Alicante Limestone. Known to architects as "Rojo Alicante," it is still actively quarried on Mount Cavarrasa, just west of the seaport city of Alicante in southeastern Spain.

The contrasting lighter stone, in shades of cream and rose with a much more fragmented look, is northern Italy's Fumane Breccia. It comes from quarries 3,000 feet above sea level on Monte Pastello, in the town of Fumane, just north of Verona. This is the same locale that produces the famously fossiliferous Rosso Ammonitico Veronese Limestone described in the site 8.15 section. The reason that the Fumane is so distinctly brecciated is that it is taken directly from a major fault zone, where the red limestone you see here as red angular blocks and chunks was ripped apart where the Earth's crust, under great tectonic stress, ruptured. The fact that the Fumane's clasts are angular rather than rounded suggests that they did not travel far from their original position before being recemented. This rock type has, in addition to copious calcite infills and darker veining, fossils of its own, at least here and there. Some of them, round brachiopod valves, reminded Italian stonemasons, who apparently were amateur ornithologists, of the pattern of a partridge's eye (*occhio di pernice*). Hence this stone is still known by its widely used trade name, "Breccia Pernice."

5.13 The Couture

909 E. Michigan Street
Scheduled for completion in 2024
Architectural Firm: RINKA
Foundation: Steel piles to bedrock
Geologic feature: Concrete

My interest in this building was kindled on a cold and cloudy day in early February 2022, when it wasn't a building at all. At that point it was simply a cofferdammed hole in the ground. There, I saw, a pile driver was hard at work, on the side of the pit closest to where I stood. Its impact hammer rose and fell on a steel pile that had been inserted into a vertical frame. Each time the hammer struck the top of the pile, there was a sharp metallic report, something between a thud and a clank, and a bone-jarring shudder that rattled up my spine. It reminded me of the tremors I'd felt at the summits of active volcanoes visited decades before.

This percussive, foundation-laying process on the Couture's site ultimately involved driving 192 such piles, each designed to bear 400 tons, approximately 140 feet below street level. These ensure that this skyscraper has a meaningful relationship—not to mention direct and intimate contact with—the Silurian dolostone bedrock. It's essentially the same method that Milwaukee architects used a century and a half ago, but with the application of twenty-first-century techniques and materials: diesel driven machinery, forged steel instead of conifer-tree timber, and anchorage not in a hardpan layer but in the Earth's deeply blanketed crust.

Now that the Couture is a fully completed building and, at time of writing, Juneau Town's newest lake-facing high-rise, we can take a lingering look at its aboveground exterior. And what we find there, mostly, is concrete, the same material that forms its entire structural framework within.

Ubiquitous in our built environment, concrete is a substance of overtly geologic origins. *Opus caementicum*, the concrete of the ancient Romans, was made, as our modern equivalent is, from two main components—mortar and aggregate. But some of their ingredients were notably different. Where we usually use Portland cement for the mortar component, the Romans relied on a mixture of lime and *pozzolana*, a yellow or reddish volcanic ash found in southern and central Italy. And their aggregates, unlike our mixtures of sand and various sizes of rock particles, often included recycled materials—pot shards and demolition debris among them. The resultant mixture was slathered down one layer after the other, almost like masonry courses, and not poured into a single, uniformly hardening mass.

Lime is the substance that runs like a thread through these different modes and traditions of making concrete. This simple but amazing chemical substance was apparently first discovered and used twelve millennia ago, in Paleolithic times. When mixed with water, it's aggressively exothermic, and generates rather than absorbs heat. It is derived, not surprisingly, from limestone—or rather from the burning of limestone and the resulting transformation of the mineral calcite, or calcium carbonate, into calcium oxide and hydroxide. In the process, carbon dioxide is liberated into the atmosphere. Indeed, in modern times, cement-making is one of the primary anthropogenic sources of this greenhouse gas, and hence a matter of serious concern to those grappling, so far incompletely and ineptly, with our looming climate catastrophe.

You can find the Couture's external concrete in the form of cladding panels prefabricated to resemble white stone. These adorn the balcony rows to produce a horizontally banded pattern.

5.14   Button Block

500 N. Water Street
Current name: Homewood Suites by Hilton, Milwaukee Downtown
Completed in 1892
Architectural firm: Crane & Barkhausen
Geologic features: Jacobsville Sandstone, Terra-Cotta

For lovers of the Richardsonian Romanesque style, this is the Pushes-All-the-Right-Buttons Block. But, as egregiously geological as it is, it has suffered from a lack of sufficiently detailed architectural documentation. Its striking upper exterior is very reminiscent of the Philadelphia Brick of the Milwaukee Club (site 5.7), but other production centers, and most notably St. Louis, produced facing brick

FIGURE 5.7. The lower stories of Juneau Town's Button Block are clad in rock-faced Jacobsville Sandstone of the cheerful "Portage Red" variety.

of the same high quality and handsome hue. Regrettably, the anonymity of its materials also extends to both the red terra-cotta ornament and the red granite entrance column; I have not been able to track down their origins. The same would be depressingly true for the unprovenanced rock-faced ashlar of the first and second floors and exposed basement were it not for the fact its distinctive color is a dead ringer for the "Portage Red" variety of the Jacobsville Sandstone. Simply no other rock type used in these parts can boast its striking tint.

The Jacobsville is Upper Michigan's component of the Lake Superior Brownstone complex, and the "Portage Red" specifically comes from the small community of Portage Entry, situated on the southeastern shore of that ruggedly beautiful and geologically fascinating peninsula, the Keweenaw. While the Jacobsville is one of the clastic sedimentary units that were deposited in the huge, 1.1-Ga-old rent in the Earth's crust known as the Midcontinent Rift, its own age is uncertain owing to its dearth of fossils. Nor does it hold other relevant clues that would allow stratigraphers to constrain its age more tightly than its current, all-too-immense span of very late Mesoproterozoic to early Cambrian. Nevertheless, it's a wonderful rock type to peer at with a hand lens, and its iron-oxide-rich sand grains are particularly inspectable near ground level along the Water Street and Clybourn Street elevations. Here you'll see why Smithsonian Institution geologist George Perkins Merrill, writing a year before this building was completed, described the "Portage Red" as "uniform in color and of excellent texture."

5.15   Mitchell Building

207 E. Michigan Street
Completed in 1876
Architect: Edward Townsend Mix
Foundation: Timber piles
Geologic features: Berea Sandstone, Hinsdale Granite, Aberdeenshire
    Granite, Chicago Terra Cotta

The Mitchell Building is a powerful demonstration of the flamboyant Second Empire school of architecture that originated in France and once dominated the streetscapes of American cities. This edifice, Milwaukee's biggest and best example of the style, did not escape the scrutiny of a correspondent for an 1878 issue of the Boston-based *American Architect*. This critic noted that the design is "perhaps too florid in treatment for a business building; but . . . it is a monument, and the designer may be excused for making its architecture verge on what Frenchmen would call the monumental." And, as it so happens, it's quite a monument to the abundance and diversity of the Earth's crust as well.

Most of the stone you see on the façade, though all too often listed as limestone, is actually Berea (buh-REE-ah) Sandstone, an Ohio architectural rock type that has remained in production from the early nineteenth century all the way to the present. Its main quarrying district, just south of Lake Erie and west of Cleveland, is ideally situated for easy shipping by water and rail to many US and Canadian cities. This fact was not lost on cost-conscious Midwestern builders, and the stone became especially popular in Chicago after the Great Fire of

FIGURE 5.8. Detail of the Mitchell Building's entrance and façade. The exterior of this Second Empire landmark isn't just busy; it's hyperactive. The foreground's dark-gray column shafts are highly polished Aberdeenshire Granite, from Scotland. Most of the rest is northern Ohio's Berea Sandstone.

1871. Nowadays, however, its finest remaining exposures are not there but in Milwaukee, where it's prominently displayed on three of the city's most magnificent buildings. This rock was also widely known and sold as the "Berea Grit," and its abrasive texture, due to its unusually angular grains, ensured its use as grindstones, too. Some 360 Ma ago its sand was deposited by *distributaries*, streams flowing through meandering canyons in a major river delta that emptied into what paleogeographers have dubbed the Ohio Bay. This was a narrow arm of the ancient epeiric sea that covered portions of the Midwest at that time.

Unfortunately, the Berea used here, originally pale gray, has over the years been painted beige, a color it at least partially retains. When it isn't subjected to such indignity this sandstone is easy to confuse with the much more common Salem Limestone—as the misidentifications of some architectural historians describing this site and others bear witness. Like the Salem, the Berea is a *freestone*: it can be safely parted in any direction. And, also like its Indiana competitor, it's relatively soft and makes a wonderful carving medium for sculptors.

Below the Berea and occupying the exposed basement level, in all of its naked glory, is one of the city's rarer crystalline building stones. This is the still unequivocally gray Hinsdale Granite. It hails from a locality of that name, actually nothing more than a minor railroad stop now, that lies in the Mesabi Range 55 miles north of Duluth. The stone was usually marketed simply as "Minnesota Granite," probably because at the time of the Mitchell Building's construction competing igneous-rock production elsewhere in that state was still a nascent industry. Soon thereafter, however, Minnesota became a major producer of both felsic and mafic rock types quarried in various communities.

Neoarchean in age, the Hinsdale is one of the oldest rock types on display in Milwaukee. It's composed mostly of medium-grained and light-toned quartz and feldspars (orthoclase, microcline, plagioclase) and the darker minerals hornblende and biotite. The hornblende is also present here and there in considerably larger chunks that quarrymen refer to as "knots." These are especially evident on the Hinsdale's most famous application, Adler and Sullivan's Auditorium Building in Chicago, erected a decade later.

One other granite is also present, but in contrast it's a foreign introduction. The building's original owner and namesake, banking and insurance magnate Alexander Mitchell, decided to honor the land of his birth by here including two pairs of massive entrance columns fabricated from Aberdeenshire Granite, a widely sought-after export of northeastern Scotland. Radiometrically dated to about 400 Ma, it's Lower Devonian and hence the second-youngest rock type on the Mitchell's façade after the Upper Devonian Berea. Given their dark bluish-gray color, these highly polished shafts may have come from the famous Rubislaw Quarry in the city of Aberdeen itself, but there were other similar varieties. The

Aberdeenshire Granite complex also includes red and salmon-pink selections of various textures, as we'll see next door at the Mackie Building.

Every bit as geologically significant, though less accessible, are the Mitchell Building's roof decorations. These, "of elaborate design and good detail," are identified in the *American Architect* article cited above as fired clay created in the works of the Chicago Terra Cotta Company. If at least some of that original ornament is still here, it's of great historical significance. Well-documented surviving examples produced by the vanguard firm that launched northeastern Illinois as a world-class producer of terra-cotta are hard to find. Much more frequently encountered in Milwaukee and elsewhere are architectural cladding and ornament fabricated by such successors as Northwestern, American, and Winkle. Chicago Terra Cotta, formed in 1869 by architect Sanford Loring and several financial backers, used Pennsylvanian-age underclays and shales from the Illinois Basin coal-mining region, which for high-quality firing were deemed distinctly superior to local, till-derived clays. So one can rightly say the Mitchell Building is decked out in materials from three different geologic periods and four geologically distinct parts of the planet.

It is also noteworthy that this grand edifice, as one of many Cream City buildings set on pile foundations, has suffered in recent times from the downtown area's lowering water table. This has exposed the Mitchell's timber underpinnings, previously submerged in anoxic groundwater, to the air and hence to wood rot. This alarming situation prompted an extensive and costly foundation-renovation project. Thanks to that, we may hope with reason that this priceless window into America's architectural past stands for many decades to come.

5.16   Mackie Building

225 E. Michigan Street
Alternative name: Chamber of Commerce Building
Completed in 1879; addition in 1880
Architect: Edward Townsend Mix
Foundation: Timber piles
Geologic features: Berea Sandstone, Hinsdale Granite, Aberdeenshire
    Granite

Constructed with most of the same materials as its gaudily attired next-door neighbor, the Mackie Building, with its relatively sparing use of ornamental detail, is more congenial to the modern eye. Here the exterior stone is deployed in the same manner, with the Neoarchean-age Hinsdale Granite on the lowest floor, and the Upper Devonian Berea Sandstone for the façade above it. And once again, the latter demonstrates its impressiveness as a carving stone. Here at least the commercial variety of Berea used was "Halderman Blue Stone," though one source lists another Berea brand name, "Amherst Stone," instead. (In the quarry trade, any gray rock with even the vaguest hint of blue tinting had that color descriptor prominently affixed to it. It's one of many examples of marketing hype found in this line of business.)

On the other hand, the granite entrance columns certainly seem different from the Mitchell Building's; for one thing, they're red rather than gray. But they, too, are a form of "Scotch Granite"—and almost certainly another variety of the Lower Devonian Aberdeenshire group. Other regions of Scotland produced granites as well, but American surveys of the building-stone industry of the time the Mackie was built suggest that the Aberdeenshires were the only ones imported into the US in any quantity in the late 1800s. The most common of their red types was quarried in Peterhead, but once again other locales also produced similar coloration.

One last similarity to the Mitchell Building is the Mackie's use of timber piles for its foundation. Interestingly, though, here the piles bearing the greatest loads were topped up to sub-basement level with Lemont-Joliet Dolostone, the same rock shipped up from Illinois that is discussed at greater length in the State Bank of Wisconsin–Bank of Milwaukee section (site 5.18).

5.17   Loyalty Building

611 N. Broadway
Original name: Northwestern Mutual Life Insurance Building; now Hilton Garden Inn Milwaukee Downtown
Completed in 1886
Architect: Solon S. Beman
Foundation: Timber piles
Geologic features: Vinalhaven Granite, Hallowell Granite, Salem Limestone

Whatever the weather or my mood, I always approach this building with extreme positive prejudice. It became, immediately and viscerally, one of my architectural favorites the first time I saw it. For one thing it is, like the nearby Federal Building (site 5.6), a model of just how effective the often darkly clad Richardsonian Romanesque style can be when expressed instead in light gray, buff, and even a hint of pink. For another, there's its wonderfully laid out layer-cake composition, so remindful to the geologic eye of the sequence of horizontal rock strata in eastern Wisconsin. Then again, it's the perfect place to compare and contrast two of the great Paleozoic-era Maine granites that in the heyday of their production were especially favored by nineteenth-century designers.

The strata analogy cited above is particularly apropos here because, like undeformed layers of prairie rock, the oldest unit is at the bottom of the Loyalty Building; its youngest obligingly sits at the top. The exposed basement story is clad in rugged blocks of Vinalhaven Granite, quarried on the eponymous island at the mouth of Maine's Penobscot Bay. Its various trade names included "Fox Island Granite" and "Duchane Hill Granite." Though it was long listed as Devonian in age, more modern sources, including the most recently published geologic map of the state, make it Silurian instead. Regardless of its exact age, it comes from the Coastal Maine Magmatic Province, a vast assortment of plutons that intruded the upper crust after Ganderia, a wandering terrane that now makes up Maine's windswept seacoast, collided with the coast of ancestral North America.

Such rock-faced ashlar as this is not all that conducive to the close examination of its mineral constituents. Nevertheless, even at several paces you should be able to detect a faint trace of pink provided by this coarse-grained selection's

FIGURE 5.9.   The Loyalty Building is a finely fashioned layer cake of igneous and sedimentary rock types: Maine's Vinalhaven and Hallowell Granites, and Indiana's Salem Limestone.

orthoclase content. This alkali feldspar is complemented by white oligoclase, a plagioclase feldspar, as well as gray quartz and black biotite mica. The full visual impact of this stone is more apparent where it has been highly polished: the group of small, squat column shafts that flank the main, Broadway entrance above the basement level. According to an 1885 description in the *Inland Architect*, these were taken specifically from Vinalhaven Island's Duchane Hill quarry. Here we have an excellent demonstration of how substantially different in tint and tone a single rock type can appear depending on the finish it's been given.

The next story up features the lighter and distinctly finer-grained Hallowell Granite. It too was a highly regarded Maine building stone, but one quarried inland about 50 miles to the west-northwest of Vinalhaven. A scan of the architectural-journal ads of the 1880s reveals just how well marketed the Hallowell was. And it was also a popular choice in the monumental trade for cemetery statues and sarcophagi—as can be seen in Chicago's Graceland Cemetery, where it was used to good effect in the elaborate memorial to George Pullman, one of America's wealthiest and most notorious robber barons.

The Hallowell, unequivocally Middle Devonian in age, formed as a body of magma that solidified far belowground during the Acadian Orogeny. This major mountain-building event occurred when what is now New England merged with yet another peripatetic landmass, the microcontinent Avalonia. Since the Hallowell is here displayed in both dressed-face (smooth but unpolished) and rock-faced form, you can fairly easily distinguish its crystals: porphyritic, white and faintly bluish feldspars—mostly oligoclase and orthoclase—as well as glassy gray quartz and two micas, black biotite and silvery muscovite.

Rising above the Hallowell to clothe the remaining floors is the architect's favorite limestone, the Salem. Better known to builders as "Bedford Stone" and "Indiana Limestone," this Mississippian-age sedimentary rock provides a congenial and slightly contrasting buff color. Like the granite below it, it has been skillfully carved into intricately detailed columns, capitals, and ornamental trim.

5.18   State Bank of Wisconsin–Bank of Milwaukee

210 E. Michigan Street
Alternative name: Insurance Exchange Building
Completed in 1857 (State Bank); 1858 (Bank of Milwaukee)
Architects: George Mygatt & Leonard Schmidtner (State Bank); Albert
    C. Nash (Bank of Milwaukee)
Geologic features: Lemont-Joliet Dolostone, Cottonwood Limestone

Originally two distinct buildings separated by a party wall, this composite Italian Renaissance Revival edifice, now joined internally, is yet another of this neighborhood's intriguing architectural time capsules. But it's even more than that. Finished a few years before a still relatively obscure politician named Abraham Lincoln won the Republican nomination for the presidency, it is the oldest structure in its vicinity, and therefore a priceless indicator of early stone use in this city. It should come as no surprise that its main, southern façade is largely composed of Regional Silurian Dolostone. But what's unusual is that the rock you see here was not brought in from nearby Wauwatosa, but from the Lemont-Joliet district, 100 miles to the south in the Lower Des Plaines River Valley southwest of Chicago.

Why was such a distant source chosen? Probably because the first Wauwatosa quarry, run by the Story Brothers, had just started operation in 1855 or thereabouts, and its own stone was not yet well known or preferred. The Lemont-Joliet Dolostone, on the other hand, had been extracted for local architectural use no later than the 1830s, and was then widely marketed and distributed throughout the region after the 1848 opening of the Illinois and Michigan Canal. That's why even today this rock can be found here and there in the built landscapes of other southeastern Wisconsin towns as well: at first Badger State stone producers were just not able to meet the demand. In following decades, however, they definitely were.

In its heyday the Lemont-Joliet Dolostone was usually marketed as "Athens Marble"—a seemingly shameless example of deceitful advertising that suggested, quite incorrectly, that the rock was (or at least was similar to) the world-renowned Pentelic Marble of ancient Athens. In fact, its main source, the lovely little river town of Lemont, originally bore the name of that great Greek city. Even so, the rock was not a marble by either geological or architectural definition. For one thing, it isn't metamorphic; and though it's soft and carvable, it simply can't take a high-gloss polish. In more recent times, the Lemont-Joliet and Wauwatosa Dolostones, in common with equivalents hailing from such places as Lannon, Waukesha, Racine, Valders, Cedarburg, and Fond du Lac, have been routinely misidentified by architectural historians as limestone. In all fairness, that's quite understandable. The two rock types are closely related and, if one does not have a bottle of dilute hydrochloric acid handy, hard to tell apart. Adding to the confusion is the fact that early geologists called our Silurian bedrock "magnesian limestone"; such terms as "dolostone" and its variant "dolomite" were not yet in widespread use among Midwestern scientists.

Unlike its Wauwatosa equivalent, much of the Lemont-Joliet Dolostone was taken not from the reef-bearing Racine Formation at or near the top of the Silurian stratigraphic column but from underlying beds called, in Illinois, the Sugar Run Formation. While similar to the Racine in composition, depositional environment, and light-gray color when freshly quarried, the nonreefal Sugar Run tends to weather more noticeably into a warm and buttery golden yellow, and even brown and ocher in places. Unfortunately, its susceptibility to alteration by the elements also means that it frequently spalls, peels, and crumbles more easily as well. Consequently, many Chicago landmarks made of it—most notably the South Side's Union Stockyards Gate—shed all too readily.

And here too the Lemont-Joliet has suffered significant deterioration, especially in its carved ornamental details. Some of these details, including those at the entrance, have been replaced by duplicates made of Kansas's Cottonwood Limestone. This rock, like the Lemont-Joliet, is a chemically precipitated sedimentary type. Officially classified as the Cottonwood Member of the Beattie Limestone Formation, it's further parsed by petrologists as a *wackestone*, a limestone that formed from carbonate mud containing some grains of fine sand as well. The Cottonwood traces its origin to a distinctly different chapter of geologic history, the early portion of the Permian period, approximately 280 Ma ago. The Permian was the final major subdivision of the Paleozoic era, and at its end witnessed the most horrendous mass extinction event in the long saga of multicellular life. But that nightmarish episode of global warming and vast outpourings of lava occurred some 40 Ma after this stone was deposited in the shallow saltwater of a narrow seaway covering what was then western Pangaea, and is now the Great Plains.

5.19   First National Bank Building

735 N. Water Street
Current name: The CityCenter at 735
Completed in 1914
Architectural firm: D. H. Burnham & Company
Foundation: Rock caissons
Geologic feature: Milford Granite

Not completed until two years after head architect Burnham's death, this site reflects his later-career commitment to the Beaux Arts ideal of neoclassical grandeur. And given the fact it was designed by a Chicago firm, it's not surprising that the First National Bank Building is anchored in its riverside setting not with Milwaukee's standard method, piles, but with caissons more characteristic of Windy City practice. One modern news account says that the caissons reach bedrock at 60 feet down, but if so, that's surprising—the old Market Square well only one block away didn't hit the uppermost Silurian dolostone beds short of 170 feet below surface. Perhaps the cited 60 should be 160?

The First National Bank's suite of exterior building materials includes pale-brown facing brick and terra-cotta in the upper stories, but neither has a documented source. The same could be said of the pilasters, spandrels, and cladding of the first three floors, were they not all made of a beautiful igneous rock type that's so distinctive to those familiar with it that it can often be identified even without provenance. This stone is the Milford Granite, quarried in and around the town of the same name in southeastern Massachusetts. My almost-certain assignment is bolstered by the fact the Milford was a favored choice of Burnham near the end of his life; it also adorns his magnificent 1911 Peoples Gas Building on Chicago's South Michigan Avenue. Of all pink granites, it's the classiest and the one with the most noticeable salmon tint. Neoproterozoic in age, it dates to approximately 630 Ma. Its parent magma was intruded into the crust of the microcontinent of Avalonia roughly 250 Ma before that far-wandering landmass collided with what is now New England and Maritime Canada.

The Milford Granite has also been widely employed in the monumental trade for cemetery monuments. On the boundary between medium- and coarse-grained in texture, it's a striking blend of pale-pink alkali feldspars (orthoclase and microcline), white plagioclase feldspar (albite), blue-gray quartz, and black biotite. So highly regarded was this stone in the late nineteenth and early twentieth centuries that the Northwestern Terra Cotta Company produced glazed, fired-clay cladding that duplicated its appearance with amazing accuracy.

5.20   John W. Mariner Building

411 E. Mason Street
Current name: Hotel Metro
Completed in 1937
Architectural firm: Eschweiler & Eschweiler
Geologic feature: Salem Limestone

Some rock types almost completely disappear with the passing of a particular architectural style—the Lake Superior Brownstone varieties are a good example—only to make brief appearances, if any, later on. But if there is one transgenerational building stone par excellence, it's the redoubtable Salem Limestone. Just in Milwaukee alone one can find its application to every design from the Richardsonian Romanesque to modernism and no doubt beyond. Still, were I pressed to single out one era in which it was used most systematically and effectively, I'd have to say it's the 1920s and 1930s, during the flowering of Art Deco and its more streamlined and less ornate offshoot, Art Moderne.

Along with the Exton Apartments (site 8.8), the Mariner Building is one of this city's most effective statements of the latter, later style. With its two principal elevations clothed in the Salem from top to bottom, it presents a smoothly rounded corner at Mason and Milwaukee. Carved detail, elegant in its simplicity, exists. But you have to hunt for it.

Not all in this expression of the Art Moderne is modern. The stone that so unobtrusively forms its buff and uniform exterior is in fact an emblem of the ancient. Across the street, it seems a geometric abstraction; when seen up close through a magnifying lens it's swimming with myriad signs of former life. These remains—everything from mote-sized forams to bits and chunks of larger invertebrates—are reminders of a tide-tossed world of 340 Ma ago.

Don't just imagine this Mississippian-subperiod scene. Don't just visualize its teeming inhuman biome set in the shoals and shallows of an equatorial sea. For one blessed minute, actually *be there*. In this earlier time the ramifying branches of evolution could go many ways and any, none of them guaranteeing arrival at our version of the present. This building, this stone tooled by artful great apes, this whole wonderful city, are the unlikely outcomes of a million unexpected turns and frightful, fruitful mistakes.

5.21 George Watts & Son Building

761 N. Jefferson Street
Completed in 1925
Architectural firm: Martin Tullgren & Associates
Geologic feature: American Terra Cotta

Unquestionably one of Milwaukee's finest shrines to the artistry of burnt clay, the Watts Building sports a style variously described as Mediterranean Revival, Moorish Revival, and Spanish Colonial Revival. However one parses it, the point is that this fanciful and highly ornamented mode was another favorite of the 1920s—especially in movie palaces and other entertainment-oriented establishments decked out in terra-cotta, a medium that's especially amenable to intricate designs and flights of architectural whimsy.

The cladding here is raked-faced and ingeniously clumpy and lumpy, perhaps to suggest adobe, and comes in beige and deeper tan. It was manufactured in McHenry County, northeastern Illinois's closest approach to rural Wisconsin both in geography and dairy-centric culture. It was there, in what was then the hamlet of Terra Cotta—now a part of the Crystal Lake sprawlopolis—that the American works was located. It stood atop a thick deposit of Pleistocene till of the Lemont Formation. This blanket of unsorted glacial sediments was laid down

FIGURE 5.10. The Cream City's finest terra-cotta confection, the Watts Building has a wealth of ornamental detail that's a hoot to scrutinize. Just right of center, the herring gull cited in the accompanying text strikes a dignified pose amid the fiddly bits of the cornice.

by that epoch's final ice sheet about 20 ka ago. The ready source of clay, coupled with the region's lower labor costs, enabled the American Terra Cotta Company to compete successfully with the Midwest's industry giant, the Chicago-based Northwestern firm. American, too, had a corporate office in the big city, but its highly skilled artisans did all their creative work amid the Holsteins.

I've never quite been sure, given all the remarkable designs I've seen in stone and bronze, why it's the sculpted shapes in terra-cotta that most fascinate me and leave me rapt in admiration. Here you'll overdose on the repeating patterns and processions of winged griffins, urns, medallions, shields, twisted columns, palmette acroteria, Corinthian pilasters festooned with vines, and finials that to my eye look like heads of garlic. Once, while I was photographing the Watts Building's eastern façade, a herring gull was perched atop the cornice. For ten minutes or so I thought it was part of the design. But then it stretched its wings, uttered a contemptuous *ha-ha-ha*, and launched itself lakeward.

5.22 Matthew Keenan Townhouse

777 N. Jefferson Street
Completed in 1860

Architect: Edward Townsend Mix
Geologic feature: Cream City Brick

The Keenan Townhouse provides the urban explorer with an excellent introduction to high-quality Cream City Brick as it appeared in its original, pristine
condition. It also has quoins and window trim of what historians are always
obliged to call "limestone." Given the building's relatively early completion year,
it's unclear whether this really means Lemont-Joliet Dolostone, its locally derived
Wauwatosa equivalent, or maybe even something else. The sources are silent.

In 1984 a fire destroyed this building's interior, but fortunately for all fans
of Milwaukee architecture the decision was made not to tear down its smoke-
stained but still-intact outer walls. The structure was rebuilt within and reinforced with a thoroughly modern steel frame, while its preexisting exterior was
respectfully refurbished. As a result, we can now see how this residence appeared
to Milwaukeeans strolling down Jefferson Street just before the start of the American Civil War. And this is definitely the place to more deeply explore the history
and geology of its fabulous building material.

It seems that the production of Cream City Brick began in the year 1835, and
soon became one of the area's major industries. As was the case in Philadelphia,
St. Louis, and other major urban centers famed for their brick, production yards
here were established wherever an ample supply of suitable clay was accessible.
And that included many places in the floodplains of the county's three major
streams, in both Milwaukee and Wauwatosa. The very first of these yards, run by
the Olin Brothers, was situated at the intersection of Huron (now Clybourn) and
Water Streets, on the eastern bank of the Milwaukee River, essentially where the
Button Block (site 5.14) now stands. By 1880, the largest brickyard of its time,
located on the south side of the Menomonee River Valley at 13th Street, was
turning out more than fifteen million bricks a year. Produced in both common
and facing forms, Cream City Brick was used extensively here, as is still obvious
today, and was shipped to other American cities from at least Minneapolis to
New York. Much closer to home, it can be found in Kenosha's historic Southport
Lighthouse, Waukegan's Ballentine House, and Old St. Patrick's Church, on Chicago's Near West Side. Still, to understand its immensely positive visual impact,
one must see it en masse, where it has been the one most distinctive and repeated
element in Milwaukee's architectural legacy.

Sadly, the market for Cream City Brick fell into eclipse by the beginning of
the twentieth century and suffered complete extinction by the end of the 1920s.
It had fared badly in pricing wars with Chicago brick producers, but another
major reason for its demise lay in a tale often repeated for building materials
once admired as the epitome of taste and style. The fickleness of human aesthetic

judgment, so mutable from fad to fad and generation to generation, also did much to bring it low. As the *Architectural Record* noted dismissively in 1905,

> The cream-colored brick in which the city at one time took especial pride has fallen into disfavor, and justly enough, for in color it is thin and cold, with no value except perhaps in contrast with new-fallen snow. It is particularly ugly in its cheap, rough grades, as used in blank party walls and on inferior buildings, here it turns, when stained with soot and weather, to a dreary, sickly, streaked gray—as utterly a forlorn building material as can be imagined. For all the better class of work nowadays the brown, red or pink brick of other localities is imported.

But why is Cream City Brick, which happily survives at many sites described in this book, so characteristically light yellow? The answer lies in its geochemistry. The lacustrine and glacial clays from which it was made contain a superabundance of calcium and magnesium, a gift at least partially derived from our underlying dolostone bedrock. When Wisconsin geologist Ernest Robertson Buckley analyzed the chemical composition of brick clays in Milwaukee and such other places as Madison and selected locales in Alabama, Maryland, and Colorado, he discovered that the iron concentration remained pretty much the same, ranging from just 1.5 to 4.4 percent. But the calcium varied widely: in Cream City clays, it was 14 or 15 percent, and consistently 1 percent or less elsewhere. And magnesium had a similar pattern, with 8 percent here and 1 percent or lower everywhere else. What all of this indicates is that the yellow Cream City Brick color is not the result of a dearth of red-producing iron, but of the unusually high concentration of calcium and magnesium, which tends to neutralize the iron's effect. That said, Milwaukee County brickyards were also capable of turning out bricks of other colors, ranging from bone white to salmon and even a muddy green. It was partially a matter of which clay layer or layers were used—their chemistries varied a bit—and also of the temperatures the clays were subjected to.

Much more can be learned about this subject by reading Andrew Charles Stern's 2015 history of Cream City Brick, which is well-researched, clearly written, and highly recommended. Also of interest is Buckley's 1901 *The Clays and Clay Industries of Wisconsin*. Both accounts are listed in the Selected Bibliography at the end of this book.

5.23   Wisconsin Consistory

790 North Van Buren Street
Alternative names: Scottish Rite Cathedral; currently the Humphrey
   Scottish Rite Masonic Center

Completed as the Plymouth Congregational Church in 1889; converted
to Masonic Order facility in 1912; extensively rebuilt in 1937
Architects: Edward Townsend Mix (1889); Leenhouts & Guthrie
(1912); Herbert Tullgren (1937)
Geologic features: Wauwatosa Dolostone, Salem Limestone, Vermont
Mixed Slate

This building of rather offbeat mien began as a Richardsonian Romanesque
church and over the course of decades was transmogrified into what is now
mostly an Art Moderne design. Still, given the survival of such Richardsonian
fingerprints as a rustic slate-tile roof, hip-roofed dormers, a rock-faced stone
base, and a circular tower with a conical top, I'm dubbing it Romanesque Moderne instead. Regrettably, this remarkably clever term is unlikely to appear in
any forthcoming treatise on architecture thanks to its lack of duplicability, and
because the aesthetic judgments of geologists should never be trusted.

With the completion of its 1937 major makeover, all that was left of the lot's
preceding occupant, the Plymouth Congregational Church, was the exposed
basement level. Though I gather its exact identity has never been documented,
the rock here is what I take to be a good exposure of Silurian-period Wauwatosa Dolostone, which some historians prefer to call "Milwaukee County Limestone." Above it, in ascending grooved pilasters and some intriguing sculpture,
is the ever-adaptable, Mississippian-subperiod Salem ("Bedford," "Indiana")
Limestone.

Since both these stone types are well represented elsewhere in this book, our
attention is best directed to the roof itself. It's composed of Vermont Mixed
Slate, whose assortment of green, gray, and purple tones produces a pleasant,
mottled effect. Quarried just east of the Vermont–New York border, this foliated metamorphic rock has long been prized for its *fissility*—it splits readily into
thin sections. While roofing and flagging are slate's best-known architectural
applications, it also has long been used in other roles as well—for example, for
billiard-table tops and, in case anyone still remembers them, classroom chalk-
boards in the days when no teacher ever finished a lecture without first getting a
little coated with white calcareous dust.

While other rock types usually have only one particular point of origin on
the geologic time scale, metamorphic rocks in fact have two: the age of formation of the *protolith* or parent rock, and the age of its subsequent mutation by
heat or pressure. In the case of Vermont Mixed Slate, it's composed of rock taken
from at least three different beds whose varying colors indicate slightly different
depositional environments. The purple came into being where the iron mineral hematite was prevalent; the green, where chlorite was instead; and the gray
or black, where either sericite or abundant carbon compounds predominated.

But all began as marine shales in the deeper water of a continental shelf during the Neoproterozoic era. Later, in the Cambrian and Ordovician periods, these sedimentary units were plowed up by an approaching arc of volcanic islands and shoved some 60 miles onto the edge of Laurentia, the early Paleozoic version of North America. This was part of the mountain-building event known as the Taconic Orogeny.

5.24   Cudahy Tower

925 E. Wells Street
Completed in 1909 (southern, Buena Vista Flats section); 1929
    (northern, tower section)
Architectural firms: Ferry & Clas (1909); Holabird & Root (1929)
Geologic features: Atlantic Terra Cotta, Northwestern Terra Cotta,
    Glazed Brick

To gaze at this bright and gleaming mass is to see at once the virtues of terra-cotta and its close cousin, glazed brick. These two lightweight and relatively inexpensive cladding materials, easily cleaned, permitted the existence of overtly white buildings in polluted cities, where stone quickly became stained with soot and airborne particulates. Even the Salem Limestone, so highly regarded for its dependability, couldn't compete in that department. As architect Thomas Tallmadge wrote in 1941, "it is a magnificent stone to carve but dirt sticks to it like a poor relation."

The lower, southern, Buena Vista Flats portion of this complex has the special distinction of being dressed in Atlantic Terra Cotta. Unlike the later tower section, which instead features Chicago's Northwestern Terra Cotta made from Illinois Basin underclays and shales, it was fabricated from material mined from the Upper Cretaceous Raritan Formation of New Jersey. This extensive sequence of clay, sand, and silt, a prominent geologic component of the Atlantic Coastal Plain Province, was deposited on the passive margin of North America after it parted from Africa during the breakup of the Pangaean supercontinent. And never in the 90 Ma of the Raritan's existence did it turn to rock. In the Perth Amboy area of New Jersey, where the Atlantic works was based, the Raritan's clay provided a superb base medium for high-quality terra-cotta. As a result of mergers with other New Jersey and Staten Island competitors, Atlantic became the colossus of its industry—and one that proudly claimed in trade-journal ads of the early twentieth century to be the largest manufacturer of architectural terra-cotta in the entire world.

FIGURE 5.11.   Looming up out of the drifting lake fog are the two sections of the Cudahy Tower. In the foreground, the 1909 Buena Vista Flats, clad in white Atlantic Terra Cotta; behind and higher, the 1929 addition, with a matching exterior fabricated instead by the Northwestern works.

5.25   University Club Tower

825 N. Prospect Avenue
Completed in 2007
Architectural firm: Skidmore, Owings & Merrill
Foundation: Caissons
Geologic feature: Concrete

This high-rise haven for Milwaukee's moneyed elite has just one claim to geologic or even general interest: it rests on caissons rather than the more usual pile foundation. Seen at close hand, the bland exterior does not really demonstrate concrete's great ornamental potential, though white horizontal bands of something resembling ceramic or porcelain tile have been added as a token gesture on its lower section.

On one of my research trips, I brazenly poked my head into the first-floor lobby, having read that it contains some sort of burgundy-tinted German stone I'd never come across. There I was angrily accosted by the first undiplomatic concierge I've met in the long record of my travels. I gather he'd been trained by his overlords to rebuff even the most fleeting incursions of the poor, curious, or unfortunate. And rebuff me he did, but not before I'd copped a good look at the enigmatic stone in question. One day I'll come across its true identity, or you will. Regardless, it won't be accessible to any reader of this book who hasn't made it to the One Percent.

5.26   St. John the Evangelist Catholic Cathedral

802 N. Jackson Street
Originally completed in 1847; tower replacement completed in 1893; partial reconstruction after a 1935 fire completed in 1943
Architects: Victor Schulte (1847); George B. Ferry (1893); William R. Perry (1943)
Geologic feature: Cream City Brick

St. John's is one of the city's most splendid examples of Zopfstil, a form of German neoclassicism on display at various Milwaukee sites. It's also a great place to understand just how monumental and imposing Cream City Brick can look in an ecclesiastical setting. To make sure the exterior of the new tower with its more Baroque design would at least match the original section in texture and color, architect Ferry used Cream City Brick taken from houses recently torn down—an early example of the canny utilization of recycled building materials.

There are also two unknown stone types present on the façade, which, given the cathedral's complicated restoration history, may be of different years of installation. Forming the water table at the bottom is a gray, coarse-grained granitoid; just above it and serving as the pilaster bases is a buff clastic sedimentary rock type that is badly spalled in spots. Most likely, it's one of two Ohio selections: Berea Sandstone or Buena Vista Siltstone.

5.27   Old St. Mary's Catholic Church

844 N. Broadway
Originally completed in 1847; a substantial enlargement completed in 1867
Architect: Victor Schulte (1847 and 1867)
Geologic feature: Cream City Brick

This grand old Milwaukee landmark shares three things with St. John's Cathedral: the same original architect, the same general adherence to the Zopfstil

FIGURE 5.12.   Cream City Brick has the peculiar property of being handsome both clean and grimy. Old St. Mary's Church, just about the best example of the latter state, gains somber grace from the soot its surface accumulated before the Clean Air Act took effect. Were this well-sited house of worship cleaned now, it would lose half its visual impact and all its darkling beauty.

design ethic, and the same reliance on Cream City Brick. But what a difference it presents at first glance, at least at time of writing. The coloration of the individual bricks, despite the occasional remnant yellow, is mostly dark gray to flat black. This of course is the legacy of many decades' exposure to the Industrial Revolution, to a society that derives its wealth and power and squalor from the manic consumption of bituminous coal and petroleum. It's as though this one bespired building displays both our highest aspirations and grimiest byproducts.

At this point I offer a heresy. While no one could fail to regret the various deleterious effects of air pollution—here duly noted and certainly not extolled—I have to say I find dirty Cream City Brick, especially on nineteenth-century churches, at least as aesthetically pleasing as its pristine form. It confers, as the ivied walls of academe are supposed to, a sense of venerability verging on timelessness. I know this building was yellow and should be yellow, but also subconsciously appreciate that it has been cured and tempered in the oven of time. And

I consider it one of the chief architectural glories of Milwaukee and other cities of southeastern Wisconsin that there remain old cream-brick churches like this one that show they've aged well and visibly. Once merely cheerful, they now darkly point to heaven. It's good to see them as they are, before well-meaning restorationists attempt to erase what really can't be—the passage of the years.

5.28   City Hall

200 E. Wells Street
Completed in 1895
Architectural firm: Henry C. Koch & Company
Foundation: Originally wooden piles; now reinforced with steel micropiles
Geologic features: Wauwatosa Dolostone, Berea Sandstone, Holston
    Limestone, St. Louis Brick, Winkle Terra Cotta, Boston Valley Terra
    Cotta, and Gladding, McBean Terra Cotta

An hour spent with this building should be enough to convince you that it is the most marvelous city hall on the planet. Architecturally striking, justly a source of civic pride, it's also teeming with geologic details.

Starting with the foundation, we find that piles, Milwaukee's most common stabilization method, were used on a grand scale. Considering that this site formerly was river-bottom wetland, it must have been no easy task to anchor this massive, oddly shaped building that in plan view resembles a slender wedge of pumpkin pie. (Its narrowly triangular footprint was necessitated by the polygonal block on which it stands.) For twelve decades this 50,000-ton behemoth was upheld by some 2,600 piles, each 25 feet long, fabricated from the trunks of *Pinus strobus*, the Eastern White Pine. These poles were capped with courses of heavy ashlar, but it's unclear whether it consisted of Ordovician Platteville-Galena Dolostone, quarried in Duck Creek, Wisconsin, or Silurian Wauwatosa Dolostone, the rock that supposedly forms the basal, underground portion of the exterior walls. The City Hall utilization of Duck Creek stone was briefly mentioned by geologist Ernest Robertson Buckley in his 1898 report, but it can't be accounted for anywhere else in the structure, if later descriptions of the Wauwatosa basement walls are correct.

As long ago as the 1950s there was evidence that some of the timber shafts were beginning to rot, though still bathed in water provided by the building's subsurface hydration system. And by the dawn of the second decade of the twenty-first century it was all too obvious that the structure was settling differentially and needed serious foundation reinforcement. The method selected was an ingenious one employing "micropiles," sections of steel piping driven down 75 feet into the substrate muck, next to the existing pinewood piles and the piers that rose upward from them. The piers were then encased in rebar-strengthened concrete. When the hydration system is disconnected the wood piles will continue to decay, but the load of the building will be laterally transferred to the network of adjacent micropiles. The total cost of the project was reportedly over fifty million dollars, but it was well worth it to save this exceptional edifice.

The Wauwatosa Dolostone that has played a crucial structural role underground may be out of public view, but there are plenty of other building materials amply displayed in plain sight. The best way to start your geological exploration is by examining the building's matchless exterior. At its bottom runs the water table, a canny design feature that deflects rain dripping from above away from the base, and also prevents the more vulnerable main cladding stone above it from direct contact with the ground and its destructive substances. It's made of an unidentified coarse-grained granitoid rock. While it contains a healthy percentage of dark, mafic minerals, it nevertheless has an overall light-pinkish-gray color, and it's glaringly obvious that it is not, even by the loosest definition, the "black granite" that many architectural historians have made it. Apparently some august but egregiously shortsighted authority first misidentified it, and then a procession of other archi-aesthetes just parroted the party line without taking the time to really look at the striking rock that encircles the entire building. And speaking of mysterious light-toned granitoids, also make sure you note the hefty first-floor column shafts, of uncertain origin. They and the water table point out a depressingly common pattern that often confronts the urban geologist: on great buildings richly bestowed with a large assortment of building materials, some just can't be conclusively identified.

Fortunately, a fair number of commentators have correctly named the buff-toned sedimentary rock variety just above the water table. This is the same Ohio-quarried Berea Sandstone that adorns the older Mitchell and Mackie Buildings (sites 5.15 and 5.16). Here it forms the cladding, massive semicircular arches, intricately carved column capitals, and other ornamental doodads of the first two floors. Above that stretches a much greater exposure of striking pink brick—specifically the St. Louis variety, manufactured by the Hydraulic Press Brick Company as its No. 509 line. By the time of City Hall's construction, St. Louis, blessed with the perfect assortment of locally extractable raw materials, had become one of the nation's leading brick manufacturers. While its common red variety was made from *loess* (pronounced LUSS), the local Pleistocene wind-blown silt that blankets the whole urban area, fancy facing brick of the kind seen here traces its source to much older sediment, the Pennsylvanian paleosol known as the Cheltenham Clay. This bed of ancient coal-swamp underclay, which lies directly below the city's surface, was extensively mined and also used to make firebrick, sewer pipes, and other products.

Originally, City Hall's upper façades and bell tower were surfaced in some eight million St. Louis bricks. But in recent times a sizable number have been replaced by identical units fabricated elsewhere. Like stone and terra-cotta, brick-work is not immune to the ravages of our unforgiving inland climate, which reliably dishes up a savage brew of freezing and thawing precipitation, daily and seasonal temperature extremes, and caustic pollution products. But as you peer upward at the brick, do appreciate the fact that whatever its age or source, you're seeing it essentially the way it looked back in 1895. Historians writing before the cleaning and major renovation described it instead as "blackish red" owing to the buildup of many years' worth of soot and grime.

And amid the expanse of surviving and replacement brick are many examples of another fired-clay product, which, as it turns out, was also made in St. Louis. The brown terra-cotta of the spandrels, upper capitals, and balustrades was origi-nally all the work of the Winkle Terra Cotta Company, which drew its base mate-rial not from the Cheltenham bed but from another Pennsylvanian-age source, shale mined in the nearby Missouri community of Glencoe. This main ingredient was mixed with clays from elsewhere to produce the company's highly regarded decorative elements. However, by the first decade of the new millennium, much of City Hall's complement of Winkle Terra Cotta had, like the brick, deteriorated to the point that the installation of replacement units was necessary. These were crafted by two firms, New York State's Boston Valley and California's Gladding, McBean, which happen to be the nation's only remaining major producers of architectural terra-cotta. Both are justly renowned for their restoration work on older clay-clad landmarks in Milwaukee, Chicago, and other cities.

As impressive and instructive as the composition and weathering history of the façades are, I'm always soon sucked by some invisible force into the building's interior. Milwaukee residents who've had less than pleasant experiences navi-gating the bureaucracy it contains may bitterly disagree, but I've always found this space engagingly weird and nothing short of magical, in a late nineteenth-century sort of way. Its rotunda is laid out in the form of a stretched hexagon and is at first glance remarkably unmonumental. But every time I stand at one end of this unprepossessing chamber and glance up at the becolumned galleries and skylight above, the real monumental aspect suddenly asserts itself. It seems I've wandered into one of those M. C. Escher prints, where the eye and brain are bent into believing the only way to reach the bottom is to rise to the top, and vice versa: perhaps the perfect illusion for this seat of city politics. Still, was this tableau of receding narrowed replications, beckoning yet vaguely disquieting like the shadows of half-tucked raptor wings, actually intended? Was the cumulative

FIGURE 5.13. Milwaukee's most weirdly wonderful building interior, City Hall's rotunda and balconied light well feature interesting ornamental-stone selections and a sense of warped space straight out of M. C. Escher.

visual effect of all those layer-caked levels, each with their lights and doorways and railings and hovering ceilings, consciously worked out in advance? Or was it just a fortuitous result of a more innocent, less imaginative design idea? Regardless, it's a wholly distinctive public space and spectacle.

If after admiring the upward vista you lower your gaze once more to the wainscoting of the ground-floor walls, you'll there find the rock type that has probably adorned more American building interiors than any other. Known to everyone in the building trades as "Tennessee Marble," this is the Holston

Limestone, quarried in a long, narrow swath southwest to northeast of Knoxville. Ordovician in age, it formed on the continental shelf of Laurentia and was later thrust landward by an approaching volcanic island arc that subsequently collided with the mainland. Still considered a sedimentary rock despite its tectonically impressive track record, the Holston is tight-textured enough to take the same high-gloss polish that more fully metamorphosed true marbles do. It comes in various patterns and color schemes, but the light-brown variant displayed here has long been its most popular, and can be distinguished by its signature feature, *stylolites*. These are produced in carbonate strata when compressive pressure, due either to the accumulating weight of overlying deposits or to tectonic activity, cause films of soluble ions to migrate from a particular zone in a bed. Darker insoluble materials that remain then form visible wavy surfaces within the rock that, when exposed along a cut, two-dimensional surface, look like squiggly lines resembling a series of miniature lightning bolts zipping and crackling through the stone.

Also gracing the rotunda, albeit rather unobtrusively, are scroll-topped columns with stout shafts of a very handsome but so far unidentified brecciated limestone or marble. These have been given a high polish that accentuates the angular buff clasts set in a matrix of contrasting rose-pink.

5.29   Pabst Theater

144 E. Wells Street
Completed in 1895
Architect: Otto Strack
Geologic features: Salem Limestone, St. Louis Brick, Cream City
    Brick, Terra-Cotta, Cast Iron

This corner of town should be dubbed Little St. Louis. Here that city's brick defines the façades of three adjacent architectural landmarks. However, in contrast to City Hall just across Water Street, the St. Louis Brick of the Pabst Theater's front face is orange rather than pink. Whatever its color, it's always a class act, and no doubt it has helped this building get its share of rave reviews from the critics.

While it's definitely upstaged by the Missouri-born star of the show, there's also locally produced Cream City Brick. Look for it behind the scenes, so to speak, on the theater's northern wall. Terra-cotta of unknown parentage is also an effective supporting actor. It provides fancy ornamental detail, while cast iron plays a more prominent role in the handsome porch and balcony.

Two types of stone have been scripted into the cast of characters, too. The striking polished entrance columns are made of a red granitoid of unrecorded identity. As far as the cladding of the lower two stories go, some historians, including the one who drafted the nomination for the theater's inclusion on the National Register of Historic Places, claim that it's sandstone, and for that geologically ham-fisted performance they deserve a few hisses and catcalls. In fact it's Salem Limestone, the Mississippian-age biocalcarenite that, under its stage names "Bedford Stone" and "Indiana Limestone," is the longest-running show in the American building-stone biz. You should be able to spot some of its tiny fossil fragments locked in their calcite matrix.

5.30   Oneida Street Station

108 E. Wells Street
Completed in 1900
Architect: Herman J. Esser
Geologic features: Salem Limestone, St. Louis Brick, Terra-Cotta, Bronze

The front end of what was once the local coal-burning power station, this neoclassical gem bears a name that reminds us that what is now Wells was once Oneida Street. The structure harmonizes so well with the Pabst Theater next door because it too is clothed in Salem Limestone, St. Louis Brick, and complementary terra-cotta. It also sports what I presume is a bronze address plate at its southeastern corner.

The most important thing about the building is that it offers the urban geologist a special opportunity to closely examine the orange brick that extends downward on the façade to eye level. In it swim iron-rich specks of maroon, a trademark of some of the premium varieties of St. Louis Brick. To produce this subtly pleasing effect, the clay, once the soil in which Pangaean coal-swamp vegetation grew, had to be fired at an unusually high kiln temperature. This allowed the iron content to migrate to the surface. Note how the terra-cotta trim is similarly spotted.

And besides the bronze address plate on the corner, there's the Salem Limestone base. Lacking the protection of a granite water table, it has in places suffered very badly from contact with wintertime meltwater laced with deicer salts. Once this porous stone becomes saturated, it begins to exfoliate badly. At time of writing, one can find spots where the comb-chiseled surface of the Salem is peeling off in large sections like layers of dead skin, to reveal a degraded core of rotten stone. No designer or building manager wants to see this, of course. But actually it's just one more spin of the grand wheel of destiny geologists call the Rock Cycle. The particles now being weathered away here once had been swept along in the tidal currents of a tropical sea. Later, they formed the lithified crust of southern Indiana. Now, having taken a spell in a structure erected by an

FIGURE 5.14. The southeastern corner of the Oneida Street Station presents an excellent inventory of geologically derived building materials. From top to bottom: iron-spotted St. Louis Brick, a bronze address plate, terra-cotta that's also spotted, and dressed-faced Salem Limestone.

animal species programmed to both build and ruin things, they return to their component compounds. Ultimately they'll be put to good use somewhere else, in another of nature's own construction projects.

5.31   Kilbourn Avenue Bridge

Kilbourn Avenue at the Milwaukee River
Completed in 1929
Architect: Charles Malig, Milwaukee Bureau of Bridges & Buildings
Geologic feature: Salem Limestone

To our growing roster of Salem Limestone applications we can now add this one, as the preferred ornamental material for Beaux Arts bridges. The Kilbourn Avenue span, in my mind Milwaukee's most beautiful, shares much in common with the one 80 miles to the south that carries Michigan Avenue over the Chicago River. Both are of double-leaf trunnion bascule design, with two sections that on opening part in the middle and, pulled by counterweights, gracefully lean back on rear-mounted axles. And both feature a quartet of dignified bridge houses decked out in the one rock type, elegant in its understated way, that's best suited to the neoclassical style. If you take a stroll across this bridge, you'll see that its balustrades are also made of Salem Limestone.

5.32   MGIC Plaza

250 E. Kilbourn Avenue
Completed in 1972
Architectural firm: Skidmore, Owings & Merrill
Geologic feature: Tivoli Travertine

An ancient Roman architect suddenly teleported to this modernist plaza would no doubt suffer a serious case of culture shock. He might even conclude he'd landed among a tribe of witless, bone-chewing barbarians—a people too unschooled in art and good order to erect the colonnades, temples, public baths, and triumphal arches that to him signal the basics of urban civilization. But then he'd spot the coffered ceilings of this building's overhangs; they might remind him of the dome of his own city's Pantheon. And better yet, he'd see stone that covers the headless piers and blank surfaces of this curiously cantilevered structure. In that off-white, banded and pitted rock he'd recognize a tried-and-true friend greeting him in this alien environment.

That rock is the Tivoli Travertine. A type of limestone quarried east of Rome that dates only to our present, Quaternary period, it's age-old nonetheless because it's been in architectural use for two millennia. (See site 5.1 for more on its history and origin.) Here it serves as the exterior cladding for both the lower building at hand and its taller companion behind.

Even if a time-traveling Roman visitor couldn't be expected to understand the design ethic expressed here, our twenty-first-century eyes should. As the inheritors of styles and cultures undreamed of in centuries past, we can recognize the grace and light-bathed openness invested in these simple shapes that stand before us. After all, this is one of Milwaukee's most successful marriages of pleasing space and savvy positioning.

5.33   Marcus Center for the Performing Arts

929 N. Water Street
Completed in 1969; major renovation completed in 1997
Architectural firms: Harry Weese & Associates (1969); Kahler Slater
    Torphy with Engberg Anderson (1997)
Geologic features: Oneota Dolostone, St. Cloud Area Granite

In addition to featuring two interesting types of rock cladding, the Marcus Center has an intriguing if troubled history of prior stone use that offers a clear example of the capricious effects of differential weathering.

For this building's original design world-renowned architect Harry Weese chose Tivoli Travertine, the same material that clothes the MGIC Plaza described in the previous section. Apparently Weese was especially fond of the Tivoli; he also used it for the Seventeenth Church of Christ, Scientist in Chicago's Loop. Completed one year before the Marcus Center, that house of worship still wears its original travertine skin with no sign of substantial deterioration, as do the MGIC buildings. In contrast, the Marcus Center's exterior stonework had developed serious problems by the mid-1990s: the cladding panels and their fasteners were in trouble, with buckling and other signs of material failure evident.

The Tivoli's poor performance at this one site was due, it seems, to the fact its sections had been cut too thin. It was an excellent example of the law of unintended consequences. Advances in Italian stonecutting technology in the mid-twentieth century had for the first time allowed quarried slabs of rock to be sliced more finely—a cost-cutting feature much appreciated by architects and builders mindful of the need to submit competitive bids. But when these thinner panels of such relatively porous rock as Tivoli Travertine were installed on structures like this one in the American Midwest and Northeast, the temperature extremes and abundant moisture that are the hallmarks of a continental climate sooner or later did them in. These unforgiving conditions lifted the lid of a Pandora's box of problems. These included microfractures and panel warping, and fastener corrosion that led to the cladding's demise. As Tim Grundl, Nancy Hubbard, and Tom Kean noted in their University of Wisconsin–Milwaukee website (see the Selected Bibliography), this is essentially the same problem encountered a few years earlier on the Windy City's colossal Aon Center, formerly known as the Standard Oil and then the Amoco Building. There similarly slim sections of premium-grade Carrara Marble had to be entirely replaced at immense cost with more thickly cut units of North Carolina's Mount Airy Granodiorite. (For more on that story, see my more extended Aon Center account in this book's companion volume, *Chicago in Stone and Clay*.)

At the Marcus Center, replacement cladding installed by 1997 came in two forms. The upper and more extensive portion is a Tivoli look-alike often known in the trade as "Winona Travertine," though in fact it isn't that type of spring-deposited limestone at all. Instead, it's one of two popular forms of Minnesota-quarried Oneota Dolostone—the paler version that is often dotted with *vugs*, or small cavities. These pits are generally not arranged in the wavy bands characteristic of the Tivoli, but depending on how the stone is cut and finished, they can be seen to be linked to a network of branchlike structures that signal the presence of trace fossils more fully described in the Eagles Club (site 6.25) description.

The Marcus Center's other replacement stone choice forms the substantial water table under the Oneota. It is another widely used Minnesota product, the "Diamond Pink" variety of the St. Cloud Area Granite family. Dating to about 1.78 Ga, it belongs to a group of intrusive, felsic igneous rocks known collectively as the East-Central Minnesota Batholith. This massive upwelling of magma, which may have been generated by plate convergence, crustal stretching due to slab rollback, or a combination of both, solidified well before reaching the surface at a time roughly midway between the Penokean and the Yavapai Orogenies. These two mountain-building events were triggered by the accretion of wandering terranes onto Laurentia's Archean core.

The St. Cloud Granite complex includes stone selections of an impressive assortment of colors and textures. The "Diamond Pink" seen here is decidedly coarse-grained and *porphyritic*, with some crystals (the pinkish-beige potassium feldspar component) noticeably larger than those of white plagioclase feldspar, black biotite mica, and glassy gray quartz.

5.34   1000 N. Water Street

Completed in 1991
Architectural firm: HKS Inc.
Geologic feature: Lac du Bonnet Quartz Monzonite

This high-rise, with its arresting metallic-pink windows and duller pink cladding, is one of Milwaukee's best examples of the postmodernist return to rampant stone use. And what a stone to use: the bulk of the exterior, above a water table of unlisted "black granite," is the Lac du Bonnet Quartz Monzonite. This striking, medium-grained felsic intrusive rock is similar to true granite but has a smaller amount of quartz (only 5 to 20 percent). It's composed primarily of black biotite and two lighter-tinted minerals, microcline (an alkali feldspar) and plagioclase feldspar. It also contains traces of zircon, magnetite, and apatite.

Neoarchean in age, the Lac du Bonnet is quarried in the southeastern portion of Canada's Manitoba Province, about 8 miles west of the town of Pinawa. It formed as part of a batholith emplaced on the western margin of the Superior Craton 2.5 to 2.7 Ga ago. This date range corresponds to the Algoman Orogeny, a series of mountain-building episodes that took place as the Superior terrane, one of the Earth's primordial sections of continental crust, grew considerably in size as it collided with other landmasses. In doing so, it became Laurentia, North

America's Proterozoic and Lower Paleozoic predecessor. It may be about 1 Ga younger than the Morton Gneiss seen at various other sites in town, but the Lac du Bonnet Quartz Monzonite is still one of Milwaukee's most ancient building stones. And it's an emissary from a remote time of microcontinents careening like carnival bumper-cars into early island arcs and other, mysterious terranes.

5.35 German-English Academy

1020 N. Broadway
Current name: Direct Supply Innovation and Technology Center, Milwaukee School of Engineering
Completed in 1892
Architectural firm: Crane and Barkhausen
Geologic features: Wauwatosa Dolostone, Cream City Brick, Terra-Cotta

To revisit the terminology I used in the site 5.8 description, the German-English Academy is one of the city's best exemplars of the magical synthesis that is the Third Version of the Classic Milwaukee Formula. Magical, perhaps, but the bright-toned combination of locally derived clay and stone products was no doubt primarily a practical measure, a sign of thrift and solid good sense. The Wauwatosa Dolostone, in rock-faced finish, provides the building's sturdy base and protects the more porous Cream City Brick above it from contact with the damp ground. But, as the perfect garnish to all this comely prudence, there is the pale-orange terra-cotta window insets and trim. *Alles in Ordnung, und doch so wunderschön.*

5.36 Blatz Brewery Complex

1101–1147 N. Broadway
Current name: The Blatz Condominiums
Completed in several sections between 1891 and 1906; later additions to the west constructed in the 1930s and 1940s
Architects: August Gunzmann and Louis Lehle
Geologic features: Wauwatosa Dolostone, Cream City Brick

Here once again is the third iteration of the Classic Milwaukee Formula, but this time the happy blend of Cream City Brick and rock-faced Wauwatosa Dolostone is applied on a massive scale. This cluster of structures with a German

Renaissance façade includes, on the eastern side of its block, stockhouses, a brew-house, a millhouse, and a boilerhouse. One interesting departure to the usual pattern of stone use is the employment of the locally quarried Silurian carbonate rock for ornamental stringcourses and trim as well as for the basal section. The Wauwatosa is present in both rock- and dressed-faced forms.

5.37   Blatz Brewing Office Building

1120 N. Broadway
Current name: Alumni Partnership Center, Milwaukee School of
    Engineering
Completed in 1890
Architect: Herman P. Schnetzky
Geologic features: Wauwatosa Dolostone, unidentified sandstone,
    unidentified gray granitoid

It's an amusing fact that while some architectural historians describing the Pabst Theater (site 5.29) mistook its Salem Limestone for sandstone, others made just the opposite mistake in their accounts of this fetching Richardsonian Roman-esque structure. What they took to be "Bedford Stone" and "Indiana Lime-stone"—trade names for the Salem—is in fact sandstone. It here serves as the arch elements and lower ashlar, though the lighter-toned rock also present in quantity is Wauwatosa Dolostone. In addition, the four polished column shafts flanking the entrance are fashioned from an unlisted gray granitoid that is fine-grained but also somewhat porphyritic.

One way a geologist can demonstrate that the Blatz Office Building's oft-misidentified sandstone is not the Salem is by applying a drop of dilute hydro-chloric acid. When that's done, there's no telltale effervescence producing carbon dioxide bubbles. Calcite-rich limestone, on the other hand, would fizz merrily away. Also absent are the Salem's characteristic fossil fragments. But that leaves the question of which sandstone this really is. Its color, weathering properties, and rather gritty feel make me think it's most likely the Ohio-quarried Berea, the same type seen on City Hall and the Mitchell and Mackie Buildings. However, it's a shame the documentation for this building is not detailed enough to either confirm or disprove my guess. Still, it should be noted that Milwaukee build-ing expert Russell Zimmermann, in his classic *Heritage Guidebook* (see Selected Bibliography), outshone his colleagues by getting its basic identification as sand-stone correct.

5.38   Grohmann Tower

233 E. Juneau Avenue
Completed in 2014
Architectural firm: Economou Partners
Foundation: Caissons
Geologic feature: Concrete

This high-rise, now a residential facility for the Milwaukee School of Engineering, is a notable example of a latter-day construction project in Milwaukee that has utilized caissons rather than piles or a shallow concrete pad for the building's foundation.

The concrete of the Grohmann Tower exterior is also worth a close look. It's essentially an anthropogenic limestone made of ingredients partly derived from natural limestone: what goes around, comes around, albeit at significant environmental expense. (Traditionally, for each pound of cement manufactured for concrete production about 0.9 pound of atmosphere-warming CO_2 is released into the atmosphere. For that reason, the concrete industry is now making well-publicized efforts to lower its carbon footprint.)

The concrete variety used here is an inoffensive off-white with a fine aggregate that includes dark flecking providing a modicum of interest, at least at close range.

5.39   Grace Evangelical Lutheran Church

1209 North Broadway
Completed in 1900
Architectural firm: Henry C. Koch & Company
Geologic features: Salem Limestone, Minnesota Brick, Terra-Cotta

One of the pleasing ironies of Milwaukee's architectural geology is that, despite the understandable dominance of locally produced Cream City Brick, there are so many other interesting types of clay products on display here as well. This lovely Gothic Revival church is one of Juneau Town's most chromatically distinctive buildings, primarily because of its ocher Minnesota Brick exterior. It provides a quirky counterpoint to the more predictable yellow of the Blatz Brewery Complex (site 5.37) just across the street.

Unfortunately, I've found no sources that specify where exactly in Minnesota this import was manufactured. Its state of origin, like Wisconsin, was a major source of brick: it boasted many types, made in many localities, from various sediment sources. But whatever part of Minnesota it really hails from, it's a build-ing material of striking aspect at any distance, from half a block away to hand lens range, where it reveals another asset, an unusual granular texture that suggests coarse sandstone. No doubt its ingredients have interesting tales to tell.

Below the Minnesota Brick is a plinth of that most commonly seen American architectural rock type, the Salem ("Bedford," "Indiana") Limestone. This porous Mississippian-age grainstone has a multitude of proper and effective uses, but here it's not a good choice for a damp course that comes into direct contact with the sidewalk. While it looks admirably well maintained, it has nevertheless wicked up deicer salts that have left white efflorescence crusts on the dressed-faced ashlar surfaces. In contrast, the orange terra-cotta of the entranceway, rose-window trim, and other ornament is the perfect, slightly sunnier companion for the brick. And it's another example of this material's ability to add intricate, custom-crafted detail to the designer's palette.

5.40 Quadracci Pavilion, Milwaukee Art Museum

700 N. Art Museum Drive
Completed in 2001
Architect: Santiago Calatrava
Foundation: Shallow, continuous concrete raft
Geologic features: Athelstane Granite, Carrara Marble, Concrete

While this biomimetic building with its diurnally pulsating *brise soleil* is often lik-ened to a bird, and while architect Calatrava may have had in mind the fluttering forms of Lake Michigan sailboats, I'm programmed by my preference for prehis-tory to produce a different metaphor. This marvelously motile structure is in fact homage to Wisconsin's state fossil, the trilobite. Examples of this marine arthro-pod can be found in Milwaukee's Silurian and Devonian bedrock, and evidently here, too. Such an aptly paleontological interpretation must also have occurred to other Cream City rockhounds. After all, the beautifully engineered steel fins are a reference, subconscious or otherwise, to the ancient creature's pleated thoracic segments. The building's lake face is the cephalon, the entrance on the west is the pygidium or tailpiece. Or perhaps it's the other way around.

The fact that the Quadracci Pavilion is the city's greatest monument to its Paleozoic legacy might come as a surprise to all the MAM administrators and visitors who've had the temerity to assume it's first and foremost a showplace of art. But whether you accept my tongue-in-cheek scientific revisionism or not,

I dare you to stroll across the pedestrian bridge spanning Lincoln Drive and then enter the structure itself without falling madly in love with it. Such lines and curves and magic chambers! The organic geometry of the place is something superhuman and soaked in geologic time.

Beneath this modern masterpiece sits a foundation of a kind unusual for a large building of latter days. Unusual but not unique, because it's also found under the even younger Northwestern Mutual Tower (site 5.3). In lieu of piles or caissons a shallow concrete pad, from 2 to 4 feet thick, holds the Quadracci Pavilion in place. This method, once eschewed by nineteenth-century designers of first-generation skyscrapers because of its tendency to cause dangerous differential settling, has been given new life by improved engineering techniques and advances in materials science. It's said to be an especially good solution where the substrate is too soggy or otherwise incompetent to support substantial weight. That surely must be the case here, on a site that sits at water's edge.

According to various sources, the graceful, cable-stayed pedestrian bridge is paved with "Wisconsin granite." A vague reference indeed; this state has produced quite an array of granitoid varieties from various towns and terranes. However, in the field of geo-sleuthing, frustration is often mitigated by further exploration, and eventually I realized that the bridge's wonderfully coarse-grained stone—salt-and-peppery to slightly pink, depending on lighting conditions—resembles not only some types of Minnesota's St. Cloud Area Granite, but also the Athelstane Granite variety found both in the Federal Building and the Catholic Church of the Gesu. My guess it's the latter, because the Athelstane is a genuine Wisconsin product, quarried in far-northeastern Marinette County. If my tentative but reasonable identification of the bridge pavers is correct, then its hefty crystal constituents are black biotite and hornblende, translucent quartz, white plagioclase, and beige microcline. For more on the origin of this Paleoproterozoic igneous rock, see the site 5.6 discussion.

Indoors, in both Windhover Hall and the connecting galleries, the stone flooring is much more fully provenanced. And, unsurprisingly, it's that quintessential denizen of art museums, northern Italy's Carrara Marble. This Triassic-to-Jurassic limestone metamorphosed in the Oligocene and Miocene demonstrates its famous gleam and glimmer in polished form, even if its relative softness does make it liable to being scratched, especially at baseboard level.

Still more impressive is what has been done here with concrete. Has this workaday material's capacity for beauty and dramatic impact ever been more brilliantly displayed? The procession of asymmetrical arches, as supernaturally white as the Carrara below it, is a visual tour de force that seems to suggest some secret and sacred sequence of numbers. Each of these elegantly shaped

FIGURE 5.15. The amazing grace of molded concrete, on display in the asymmetrical arches of a connecting gallery in the Quadracci Pavilion.

concrete elements was set in specially constructed wooden molds. As Calatrava himself noted,

> Concrete to me is like a supple and malleable rock. Of all materials it is the only one that can be moulded and sculpted directly on the site, under normal conditions.
>
> Concrete, although an inexpensive material, when dealt with imaginatively can make beautiful buildings. It is, however, a difficult material and requires a great deal of expertise. And by this I mean not just

technical knowledge, but also an understanding of the inner potential for poetic expression that materials possess.

Not all of Milwaukee's concrete buildings have demonstrated this inner potential. But this one resoundingly has.

5.41   Abraham Lincoln Memorial

750 N. Lincoln Memorial Drive
Completed in 1934
Architect: Ferdinand Eiseman
Sculptor: Gaetano Cecere
Geologic features: Wausau Granite, Bronze

This statue, located in front of the War Memorial Center, offers our first look at one of Wisconsin's greatest and most widely used monumental stones, the ruddy-red Wausau Granite. It forms the 5-foot-tall pedestal on which stands the bronze rendering of Abraham Lincoln. The latter was treated to form a protective dark-brown coating of copper oxide, but in spots weathering has further proceeded to the green, copper-sulfate/copper-carbonate state.

The Wausau of the base below it is in polished finish and admirably displays its splendid coloration and medium-grained texture. Isotopically dated to 1.835 Ga, the Wausau is Paleoproterozoic in age. It formed during or just after the final phase of the Penokean Orogeny, one of the most significant chapters in the Badger State's long and complicated tectonic history. This mountain-building event was caused by the Archean-age Superior Craton's collision with two other landmasses: first, a volcanic island arc dubbed the Pembine-Wausau Terrane, and then with the more geologically mature microcontinent known as the Marshfield Terrane. Igneous petrologists classify the Wausau quarried for the building stone here as an *alkali-feldspar granite*, though there is also quartz syenite in other parts of the outcropping pluton. Whereas normal granites have a fairly equal distribution of the two groups of feldspar minerals, the Wausau has mostly the alkali rather than the plagioclase type. It's thought that most alkali-feldspar granites are *anorogenic*. In other words, they do not form during plate collisions, but rather are produced later, when the continental crust is stretching rather than being compressed.

5.42   Milwaukee County War Memorial Center

750 N. Lincoln Memorial Drive
Completed in 1957; addition in 1975

Architects: Eero Saarinen (1957); Kahler, Fitzhugh & Scott
   (1975)
Geologic features: A medley of Wisconsin igneous rock types;
   Concrete

Most of the exterior of this starkly sited modernist structure is unadorned concrete, just about as cheering to the eye as the walls of a prison exercise yard. But if you make your way down to the War Memorial Center's north-facing entrance area on the lower level, you'll see one section clad in random-course, rough-textured ashlar. Architectural histories of this building suggest that architect Saarinen here exclusively used "Wisconsin granite," but they do not specify its exact source—or sources, rather, because even a quick glance reveals that at least three or four different igneous-rock types, not all granitoid, make up this lithic montage.

The most obvious distinction between these types is color: there's a lighter red, a brownish red, a lighter gray, and a black or very dark gray as well. My best guess is that these are, respectively, Wausau Granite (described at sites 5.41 and 5.43), either a deeper-toned Wausau or Montello Granite (site 7.5), a finer-grained variety of Athelstane Granite (site 8.14), and Mellen Gabbro (also site 8.14). It's a plausible list; all of these closely resemble in tint and texture what's mounted on this face. And they're all certifiably Wisconsin-quarried. Nevertheless, these assignments are unequivocally tentative. Provenance is lacking.

5.43   Southeastern Wisconsin Vietnam Veterans Memorial

Veterans Park, North of the Milwaukee County War Memorial
   Center
Completed in 1991
Geologic feature: Wausau Granite

Powerful in its simplicity, bold in its bloody redness, this ensemble of three tall Wausau Granite shafts may just make your soul shudder with associations of the ancient and mythic—especially when it's seen against the backdrop of a clear blue sky. The site was designed with deliberate numerical symbolism. There is one lower stone post for each of the eleven years of the Vietnam War, and one bench for each of the five armed services then extant. All these are made of the Wausau, too.

FIGURE 5.16.   The Southeastern Wisconsin Vietnam Veterans Memorial. The stone used here, the Wausau Granite, is the perfect symbolic medium for its sanguinary subject.

This is the best place in the city to examine, with magnifier in hand, Wisconsin's officially designated state rock in its polished form. The minerals present in this Paleoproterozoic alkali-feldspar granite include red microcline, gray and glassy quartz, black biotite mica, and traces of white oligoclase.

6

MILWAUKEE: MENOMONEE VALLEY, KILBOURN TOWN (WESTOWN), MARQUETTE, AVENUES WEST, CONCORDIA, MIDTOWN, AND VETERANS AFFAIRS; CITY OF WAUWATOSA

6.1 US Post Office

345 W. St. Paul Avenue
Completed in 1968
Architectural firm: Miller-Waltz-Diedrich Architects & Associates
Geologic features: Lannon Dolostone, Concrete, Weathering Steel

What this Brutalist structure lacks in curb appeal it abundantly repays in geologic interest. For one thing, it's the best place in Milwaukee to find a fully accessible exposure of one of southeastern Wisconsin's most attractive and historically notable building materials, the Lannon Dolostone. It also offers an interesting manmade rock as intriguing in its own way as any naturally occurring type. And then again it's this city's great civic monument to rust. No slur is intended: when you open your mind to the cyclical way the Earth does things, you'll perceive that this ruddy byproduct of oxidation is a necessary stage in the transformation of matter, a lovely and proper result of the inexorable process of weathering.

The Lannon is here found in an unusual disposition, as large rectangular panels of varying widths mounted on the St. Paul Avenue façade. These range in color from gray to buff to ocher, and beautifully illustrate the palette of weathering tints typical of the Lannon and of our region's Silurian carbonate bedrock more generally. If you inspect the panels at close range, you'll see that they've been chemically or mechanically treated in some very unsubtle way to produce a scoured, scalloped, and crusty sort of surface. A botanist staring at this might think of liverworts and lichens clinging to a damp canyon wall; a geologist, of

119

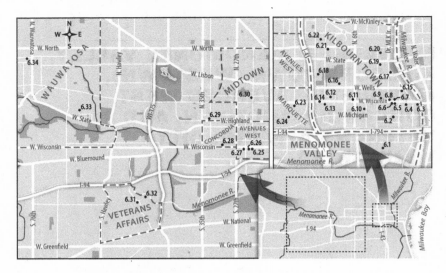

MAP 6.1.   Sites in Milwaukee's Menomonee Valley, Kilbourn Town, Marquette, Avenues West, Concordia, Midtown, and Veterans Affairs neighborhoods; and in Wauwatosa.

some wind-sculpted desert plateau. At any rate, the building's architects are to be applauded for provoking our imaginations with these abstract, otherworldly mini-terrains.

The Lannon Dolostone, taken from the uppermost beds of the Racine Formation, has a long history of use. Production in Lannon and other northeastern Waukesha County towns began as early as 1838. By the end of the nineteenth century, this rock type was primarily utilized as paving and curbing stone. However, it was also coming into vogue as an architectural stone for houses of worship, stately homes, and storefronts in many Wisconsin and Illinois communities. But unlike the quarrying of other Regional Silurian Dolostone varieties in such locales as Wauwatosa, Cedarburg, and Lemont, the Lannon operation did not ultimately fade away to the point of extinction. Today it's still going strong, though the market has largely shifted from dimension stone to veneer, landscaping rock, concrete and asphalt aggregates, riprap, and agricultural lime. But regardless of its use, it all began as interreefal carbonate mud that settled in shallow seawater some 425 Ma ago, on the western fringe of the Michigan Basin.

The rectangular columns that stand in front of the Lannon panels are full of ornamental interest, too, because their drab, light-gray matrix has been enlivened with the addition of a coarse aggregate of angular white pebbles. This makes it

FIGURE 6.1.   The adornments of Brutalism. At the US Post Office main branch, specially treated and textured panels of Lannon Dolostone are mounted under sculpted shapes of rusted weathering steel. And the pier at right is made of yet another ornamental material, pebble-aggregate concrete.

the type of ornamental concrete that I, in a desperate bid to sound scientifically hip, call calcareous anthrobreccia.

The striking stone and concrete notwithstanding, the most unusual element of the Post Office's façade is its Cor-Ten, an eye-catching form of the ubiquitous iron-carbon alloy on which our civilization is built. This weathering steel adorns the upper level. In direct contrast to stainless steel, it's specifically meant to corrode. But this is not an exercise in willful self-destruction; when it does so, it prevents further corrosion. The guiding concept in Cor-Ten is elegantly simple. Once the initial layer of rust has formed on the metal's surface, it becomes a sealant that protects the interior from further contact with oxygen, that most hyperactive and obtrusive of elements. As I always like to point out, this process is an excellent demonstration of what Earth scientists call *negative feedback*. Nothing to do with getting criticism from your boss, it refers instead to any initial condition or action that inhibits its own continuation. (Can you think of any systems in your life—perhaps in your car's engine or your home heating setup—that are other types of negative feedback?)

Though originally formulated in the 1930s for use in ore-hopper railroad cars, weathering steel eventually found its way into the otherwise relatively limited inventory of modernist ornamental materials. Its original architectural application occurred in 1964, with the construction of the John Deere headquarters complex in Moline, Illinois. Then, a year later, Chicago's high-rise Richard J. Daley Center staked its claim as its most prominent urban showplace. And not long thereafter, this main Cream City branch of the US Post Office became its first major manifestation in Milwaukee.

Most frequently rust presents us with depressing or even apocalyptic connotations. One has only to think of auto junkyards, can-strewn trash heaps, or abandoned equipment in a tumbledown farmyard. In such places it's the very essence of decay, obsolescence, and lost hope. So what a conceptual leap it really is to make it a major feature of cutting-edge building design. Still, the architect's daring feat of aesthetic role reversal makes perfect sense to the geologist. In many places on our planet the rock record displays the titanic role played by iron's transformation into iron oxide. It documents our atmosphere's evolution into a corrosive fluid rich with life-generated, life-enabling free oxygen. This can be seen most blatantly in the abundance of such red-toned sedimentary rocks as the Lake Superior Brownstones, but also in the host of iron-ore minerals—hematite, magnetite, goethite, limonite—found in soils and the crust, and exploited for uses like this by the meddling human hand.

Still, one probably unintended consequence of employing weathering steel here can be seen from the Plankinton Avenue Bridge over the Menomonee. Glancing northwestward at the Post Office's river face, you'll see how decades of dripping, rust-laden water have stained the concrete below the Cor-Ten. It looks like nature's own version of the bar-code patterns found on the thousands of boxes and envelopes that pass through this facility every day.

6.2 Public Service Building

231 W. Michigan Street
Completed in 1905
Architect: Herman J. Esser
Geologic features: Salem Limestone, Norman Brick

The exterior is a harmonious interplay between pale-buff Salem Limestone, the very widely utilized Mississippian-age sedimentary rock, and a darker tan Norman Brick. The Salem, better known in the building trades as "Bedford Stone" and "Indiana Limestone," here once again demonstrates that it's just as adaptable to the Beaux Arts style as it is elsewhere on Art Deco and Art Moderne landmarks. Note its use on this façade as rusticated ashlar, dressed-faced quoins and pilasters, and carved ornament.

Norman Brick is virtually indistinguishable from the Roman Brick described at site 6.29, especially if you don't have a ruler handy. The measurements of the Norman are usually cited in inches as 2 1/4 high by 3 5/8 deep by 11 5/8 long; Roman differs only by being, at just 1 5/8 inches high, a trifle squatter. But their actual dimensions can vary somewhat. In any event, both types are longer and lower than normal brick—a fact that makes each brick course slenderer, and that imparts a greater sense of horizontality overall.

6.3   Gimbels Department Store

101 W. Wisconsin Avenue
Current name: ASQ Center
Completed in 1925
Architect: Herman J. Esser
Foundation: Piles
Geologic feature: Terra-Cotta

Wisconsin Avenue both East and West is Milwaukee's *via mirabilis*, its Grand Boulevard of Stone and Fired Clay. If there's another urban thoroughfare on this planet that has a greater diversity of materials or a larger number of tour stops along the geologic time scale, I haven't seen it yet.

This imposing river-fronting edifice, taken with the adjoining Plankinton Arcade, constitutes its own special section, an architectural Milky Way, a heavenly shopping district. One almost suspects that in their commercial heyday these luminous white buildings were staffed with haloed, harp-toting floor managers and celestial sales clerks in flowing robes. Other terra-cotta-clad sites better demonstrate this material's endless array of colors and penchant for intricate detail. But here we see its other primary ornamental attribute: the ability to impart industrial-scale purity amid the soot and haze of urban life. The same had been attempted in earlier decades with polished white marble; but with the advent of glazed burnt clay—cheaper, more easily cleaned, and no less upscale in appearance—terra-cotta became the principal agent of chasteness.

That's not to say that terra-cotta ultimately goes unscathed in our harsh climate any more than stone can. In recent years, when the Gimbels Brothers Store was repurposed for other uses, it was discovered that more than four hundred cladding units had deteriorated to the point they needed to be replaced with replicas fabricated in California. Separately, the original foundation system of piles had to be reconditioned and enhanced. Nowadays this is an all-too-common Cream City theme. Costly and extensive restoration both aboveground and below is the price big buildings pay in daring to face, for a century or more, the perils of a wetland substrate, a varying water table, and our extremist continental climate.

6.4   Plankinton Arcade Building

161 W. Wisconsin Avenue
Completed in 1916 (original two-story arcade); five floors for offices
    added in 1925
Architectural firm: Holabird & Roche
Foundation: Piles
Geologic feature: Terra-Cotta

This westward wing of the heavenly shopping district features more copious expanses of white terra-cotta, but in this case with more elaborate designs. They're described by architectural historians as Gothic in the bottom two stories and Renaissance above. And the standard Milwaukee pile foundation system used in the Gimbels Store was also employed here. That surely presented no new challenge to the Arcade's illustrious Chicago-based designers, who had plenty of experience with both that method and caisson stabilization in their various hometown projects.

6.5   Kresge Building

215 W. Wisconsin Avenue
Now simply known by its address
Completed in 1930
Architect: J. E. Sexton
Geologic feature: Salem Limestone

As the Mariner Building and the Exton Apartments (sites 5.20 and 8.8) show, purveyors of the Art Deco and Art Moderne styles in Milwaukee had a special affinity for the Salem ("Indiana," "Bedford") Limestone. That's evident here as well, in this small but stylishly fronted store.

Quarriers and masons much appreciated the Salem's properties as a *freestone*— a rock type that can be cut or split in any direction without fear of producing wayward cracks or crumbling. Unlike less granular limestones, the Salem cannot be polished. However, it's the quintessential carving stone, and in its dressed-face form it's drab, uniform, and unassuming—just the medium to emphasize the sleek and streamlined designs of the Moderne design ethic. Make sure you take a good look at how the Salem's love of the chisel is expressed in the ornament of the second-floor façade.

If the Kresge's architect made one mistake, it was in not including a damp course of some less porous rock type to protect the Salem from the deterioration it has suffered from wicking up sidewalk water and its dissolved salts. At the time of writing, the Salem at and near grade shows all the classic signs of case-hardening stone rot, spalling, and efflorescence.

6.6 Majestic Building

231 W. Wisconsin Avenue
Completed in 1907
Architectural firm: Kirchoff & Rose
Foundation: Piles
Geologic feature: Terra-Cotta, Glazed Brick

While at first glance it might simply seem a terra-cotta tagalong of the heavenly shopping district just to the east, the Majestic was actually built well in advance of the Plankinton Arcade and Gimbels Store as we see them today. In fact, this venerable skyscraper is one of the city's most notable Beaux Arts designs.

Terra-cotta can be a remarkably convincing mimic of stone. Note how the cladding sections of the second-floor façade were molded to simulate the kind of rusticated ashlar seen on Florentine palazzi. Glazed brick is also present here at the building's summit, and, fortunately, on the western elevation as well, where it's somewhat more visible from street level. This form of fired clay has a long history; for example, the ancient Babylonians used it to sumptuously decorate their Ishtar Gate in the sixth century BCE. The modern American version was originally intended for building interiors and the reflective walls of high-rise light wells, but its ornamental potential as a façade element was finally realized at about the time the Majestic was designed.

The dark stone cladding on the ground-floor façade is most likely a later addition. While its exact identity was not recorded, it's worth some scrutiny. A coarse-textured, mafic or intermediate igneous rock of the type architects generically call "black granite," it's notable for its *labradorescence*—the ability of its perthite feldspar crystals to reflect sunlight and give its otherwise stygian appearance glittering highlights. Several ornamental-stone varieties, commercially available from the early twentieth century up to the present, are labradorescent and fit this cladding's profile. Their number includes such types as gabbro, troctolite, anorthosite, and monzonite.

6.7 Warner Building

212 W. Wisconsin Avenue
Current Name: Bradley Symphony Center
Completed in 1931
Architectural firm: Rapp & Rapp
Geologic features: Morton Gneiss, Salem Limestone

One of West Wisconsin Avenue's two exemplars of what I've dubbed the Grand Art Deco Formula—see site 6.11 for the other—the Warner Building began life as a movie palace and is now the home of one of the city's great cultural assets, the Milwaukee Symphony Orchestra. In accordance with the Art Deco Formula, the exterior's upper floors are neutral-buff Salem Limestone, and its lower reaches are clad instead in a polished and much more visually interesting rock type. The latter is the almost unimaginably ancient Morton Gneiss, the same Paleoarchean Minnesota migmatite used at the base of the Wisconsin Gas Building and the Wisconsin Tower. Here on the Warner, however, the Morton does not content itself to form a darker-toned plinth delimited to the first story or two. Instead, it invades the Salem's territory by climbing partway up the southern and eastern elevations in a pattern reminiscent of a ziggurat or a child's tower of building blocks. Its manic swirls of pink, gray, and black represent several episodes of crystallization, metamorphism, and crustal upheaval that began about 3.52 Ga ago. The sedate Salem above it is less than one-tenth as old.

6.8 Straus Building

710 N. Dr. Martin Luther King Jr. Drive
Current name: Fairfield Inn & Suites by Marriott—Milwaukee Downtown
Completed in 1924

Architectural firm: Van Ryn & De Gelleke
Geologic features: St. Cloud Area Granite, Terra-Cotta

Most of this building's exterior is clothed in light-toned terra-cotta, but to my eye the item of greater interest is the stone of the piers that extend from sidewalk to second story. This is the strikingly coarse-grained "Diamond Pink" brand of the St. Cloud Area Granite. One of a bevy of Paleoproterozoic granitoids quarried for high-grade building stone from the great East-Central Minnesota Batholith, it's the loveliest of the lot.

Look closely at the cladding panels here. They offer a superb lesson in igneous petrology. The uncommonly large feldspar-crystal size signals that the magma from which they came took a very long time to cool. We can use the geological relationship known as Bowen's Reaction Series to understand the sequence in which these minerals solidified. The first to form were the abundant black biotite grains and the small smattering of plagioclase feldspars. Then came the white and pinkish-buff alkali feldspars, half an inch or longer. They're mostly ortho-clase, but some microcline is present as well. The glassy brown quartz was last to appear, but it will also likely be the last to degrade due to its greater hardness.

This level of scientific detail shouldn't keep us from also enjoying the stone's visceral impact. In fact these two perspectives, the analytic and the aesthetic, the rational and the emotive, only give us full value when we stir them vigorously together and swallow them whole. So marvel at the chemistry and the history of those minerals, but also watch them as individual daubs of color in a pointil-list painting, where they imply both a larger pattern and the dance of the atoms within. Geologist Oliver Bowles, no slouch as a scientific descriptor, proved he had this sort of higher vision. In his 1918 survey of Minnesota's building stones he noted, "Many granites are rather dead in appearance, exhibiting the same dull, unvaried look from all viewpoints." But, as he added with a flourish of geo-poetry, "The 'Rockville' stone is made up largely of coarsely crystallized feldspars, showing brilliant cleavage faces, which on the hammered surfaces give a glitter-ing reflection. Thus, in moving past any point, there is no monotony but constant change. The rock has a tone, an individuality, like many of the finest marbles."

6.9   Henry S. Reuss Federal Plaza

310 W. Wisconsin Avenue
Completed in 1983
Architectural firm: Perkins & Will
Geologic features: Enameled Steel, Larvikite Monzonite

I realize that local wags have christened this site "the Blue Thing" and "the Milk of Magnesia Building"—the latter no doubt a gut reaction to the fact its decorative courses of enameled steel are the same color as the bottles of a well-known laxative. But I have to admit I like it, though some may find my preference hard to digest. Without this bold chromatic statement the structure would just be another of the untold thousands of soulless glass boxes now littering the Earth's surface.

And having already seen steel in its weathering form at the US Post Office, we can here appreciate another architectural application of this geologically derived alloy of iron and carbon.

To produce enameled steel panels of the kind displayed here, a specially formulated mixture of silica and fluxes is smelted into a molten mass, which is then run between water-chilled rollers that press it into sheets of solid glass. Next, the glass is pulverized into a powdery substance called *frit*, the basic ingredient used in enameling. The frit is applied to the bare metal and fired at a temperature of approximately 1,500 degrees Fahrenheit. This fuses the coating, called vitreous or porcelain enamel, directly *into* rather than *onto* the steel. The result is glossy, colorful, easily cleaned, chip- and crazing-resistant, and corrosion-free.

While the silica for the frit comes from high-quality quartz sands, the iron of the underlying steel is processed from ores nowadays extracted mostly from an unusual and unusually beautiful sedimentary rock dubbed BIF—Banded Iron Formation. Found in abundance in various places, including the iron ranges of northeastern Minnesota and Michigan's Upper Peninsula, it's made of alternating layers of red to purple chert and silver-gray hematite or magnetite. Most BIF formed in the Archean eon and the following Paleoproterozoic era, when the Earth's atmosphere had a much lower free-oxygen content than it does now.

Geologists still debate what exactly caused so much of the Earth's BIF supply to form just in that particular span of geologic history. Among currently favored hypotheses is one attributing the action of anaerobic marine bacteria that converted ferrous-iron compounds in seawater into iron oxides. Presumably this took place in foreland basins or passive continental margins where there was no volcanic activity. On the other hand, some researchers have suggested that it was in fact intense underwater volcanic activity and the eruption of high-temperature komatiite lava that caused this ore to form. Both models have much to commend them, because there are in fact two different types of BIF, the Superior and Algoma varieties, that seem to support each idea. We know that the first type did indeed form over a large area in relatively quiet basins; the latter is found interbedded with submarine lavas.

Whatever its environments of origin really were, deposition of Banded Iron Formation, especially of the Superior type, had almost completely ceased by 1.7

Ga. The world we live in now is incapable of producing it in any quantity, which makes it the ultimate nonrenewable resource. That's not to say, though, that we've come close to running out of the mineable ore deposits we've inherited from that far-distant time.

With its all-around wrapping of glass and enameled steel, the Reuss Federal Plaza exterior may at first seem completely devoid of any stone products. But at the southeastern corner of its block, at the intersection of Wisconsin and King Drive, you'll spot an open space devoted to sculptor Helaine Blumenfeld's playfully cavorting ensemble, *Family*. This is a small herd or perhaps pod of suggestive shapes, each fabricated from Norway's most famous rock export, Larvikite Monzonite. Designed to be handled, climbed upon, pushed about by visitors, and even minutely examined by hand lens–wielding rockhounds, these polished "biomorphic" figures are superb examples of Larvikite's instant eye appeal. This igneous intrusive stone contains intergrown crystals of plagioclase and alkali feldspars known as *perthite*. These reflect incoming sunlight at the feldspar boundaries and produce the metallic blue glimmers geologists call *labradorescence* or,

FIGURE 6.2.   The Larvikite Monzonite sculptures of the Reuss Federal Plaza, sparkling away on a sunny, snowy day.

alternatively, *schiller*. All this iridescent razzle-dazzle set in a dark background of pyroxene crystals makes it one of architecture's least subtle ornamental stones—one often seen adorning jewelry shop exteriors and other business establishments trying to catch the passing shopper's eye.

Larvikite Monzonite is quarried in and near the seacoast town of Larvik, some 65 miles southwest of Oslo. A type of intrusive igneous rock distinguished by its complete lack of the otherwise common mineral quartz, it dates to the beginning of the Permian period, when its parent bodies of magma were emplaced in a large rift structure known as the Oslo Graben. This great rent in the Earth's crust formed late in the Variscan Orogeny, a mountain-building phase associated with the assembly of the supercontinent Pangaea.

6.10   Schroeder Hotel

509 W. Wisconsin Avenue
Currently the Hilton Milwaukee City Center
Completed in 1928
Architectural firm: Holabird & Roche
Geologic features: Salem Limestone, St. Cloud Area Granite, Northwestern Terra Cotta, Levanto Ophicalcite

All too frequently the urban geologist finds major hotels both alluring and aggravating. While they're usually first-class showplaces of ornamental stone and clay products, the exact names and origins of these lovely and scientifically intriguing materials are often untraceable. Happily, the Schroeder, which has always been one of Milwaukee's most highly regarded venues, is a pleasant surprise in that regard. It's better documented than most.

However, while we can appreciate that this sumptuously appointed building has several identifiable items of geologic interest, the fact remains that its Art Deco exterior is hardly the most effective design of its style in the city. Its architects failed to take full advantage of either the Grand Art Deco Formula or any convincing alternative. Undistinguished reddish-brown brick more suitable on a factory wall replaces the buff Salem Limestone above the fifth floor, and it robs the latter of its most effective role as a monolithic, sleekly setback mountain. Adding to the overly fussy look is the terra-cotta crafted by Chicago's Northwestern works. This is highest-quality ornament crafted by the greatest artisans in that medium of their time, but in this instance it just compounds the sense that all the details really do not add up to anything in particular.

Fortunately, the Formula's standard basal element, darker crystalline stone, was here employed as exterior ground-story cladding. It's St. Cloud Area Granite of the "Cold Spring Pearl Pink" brand. Somewhat smaller in grain size than the very coarse "Rockville Pink" of the Straus Building two blocks to the east, it nonetheless shares with it the mineralogical composition cited in the Straus (site 6.8) section. And indeed all the St. Cloud Area Granite varieties have a common genesis as part of the Paleoproterozoic East-Central Minnesota Batholith, or ECMB. Their plutons' magma rose upward and solidified at about 1.78 Ga. Recent geochemical analysis of this massive assemblage of intrusive felsic rock types has produced a complicated origin story. Some of the granites and other granitoids present seem to have formed during the compressional, volcanic-arc phase of the subduction of the Yavapai Plate; others appear to be more characteristic of later crust thinning due to the process of slab rollback; still others have compositions that could be attributed to either. So geologists still have a lot to learn about that cryptic phase of Minnesota's prehistory.

After taking a good look at what the exterior has to offer, by all means venture into the hotel's interior. The chief glory among all the stone and plaster effects awaits you in the striking second-floor lobby, which has been carefully preserved by the current Hilton management. Here at last the Schroeder can lay claim to an Art Deco design second to none in the city. The deep-red and white-veined wainscoting, fabricated of northern Italy's Levanto Ophicalcite, serves to anchor the lighter-colored materials above it. Probably Jurassic in age, the Levanto is a breccia made up of serpentinite clasts held together by calcite. However, most serpentinites and ophicalcites are green to greenish-black in color; the unusual red tinting here is the result of the rock having been suffused with the iron-oxide mineral hematite.

The Levanto has been a highly prized "marble" since the time of the ancient Etruscans, and in medieval times and the Renaissance it was even more favored. I first discovered it in Genoa half a century ago, when I walked a few blocks from where my ship was moored and came upon San Lorenzo, Genoa's great Romanesque-and-Gothic-hybrid cathedral. There it's used both for exterior ornament and the massive columns flanking the nave. It's also on display in a much more recently erected ecclesiastical setting, Chicago's Queen of All Saints Basilica, where it can be seen in the broad base of the baptismal font. Here at the Schroeder Hotel it proves its equal applicability in a secular setting. It's arguably one of the most beautiful building stones in a city that boasts many contenders.

6.11 Wisconsin Tower

606 W. Wisconsin Avenue
Previous names: Milwaukee Tower, Mariner Tower

Completed in 1930
Architectural firm: Weary & Alford
Geologic features: Morton Gneiss, Salem Limestone, Cast Iron

This twenty-two-story expression of the Grand Art Deco Formula is yet another place to see that most unsettling stone, the Morton Gneiss, at and near ground level. The massively framed main entrance, with its cobwebby cast-iron grille, is nothing short of a tour de force, and one that shows how congenial the Morton is to being sculpted. Elsewhere on the southern and eastern elevations, each of the big cladding panels of this Paleoarchean migmatite relates its own creation myth, and, if stared at long enough, pulls the viewer into an ancient, inchoate world of vast forces and chunks of primeval matter swept along in blazing heat and ceaseless flow. I have never quite understood why purveyors of the Art Deco and Art Moderne were so often willing to unleash this representation of elemental chaos on designs otherwise devoted to balanced elegance. But the jarring contrast and weird asymmetries are utterly exhilarating. They seem to suggest that unresolved juxtapositions like this can lead to some greater harmony a step or two above the human ability to consciously comprehend.

At any rate, above this metamorphic mayhem rises the bland assurance of the Salem Limestone, the most civilized and well-behaved rock in architecture. Its calm and light-toned mass is garnished with cast-iron spandrels.

6.12   Milwaukee Central Public Library

814 W. Wisconsin Avenue
Completed in 1898
Architectural firm: Ferry & Clas
Geologic features: Salem Limestone, Terra-Cotta, Bronze, Montagnola
    Senese Marble, Scagliola

Unsurprisingly, the external neoclassical grandeur of this, Milwaukee's finest Beaux Arts building, is expressed primarily in Salem Limestone. America's most widely used architectural rock type, this Mississippian-age biocalcarenite was supplied by the Consolidated Stone Company of Bedford, Indiana. However, it seems that the Central Library's original plans specified Ohio's Berea Sandstone instead, and that only later the Salem was chosen. The switch had little impact on the building's overall appearance, though; both rock types resemble each other a great deal when seen from a distance. As you scan this vast assemblage of Corinthian pilasters and columns, carved ornament, and ashlar variously rusticated and comb-chiseled, make sure you take a good look at the two eagles perched atop large orbs over the entrance. Curiously, their bodies are made of terra-cotta but their outspread wings are bronze, as is evident from their greenish, copper-salts patina. It's a chimerical combination of building materials not often seen.

As imposing as its outer form is, the Central Library's most impressive feature, both architecturally and geologically, lies just indoors. The lobby, far from being a mere entryway or vestibule, is in fact a massive and lavishly decorated two-story rotunda topped with a coffered dome remindful of Rome's Pantheon. In one sense, however, all this appearance of opulence is somewhat deceiving. What at first glance seems to be the absolute and profligate use of an exotic, cream-to-butterscotch-colored stone turns out to be largely a case of artful trickery. Although some components you see here are indeed fashioned from Italy's

FIGURE 6.3. While the magnificent rotunda of the Milwaukee Central Public Library does feature some imported Montagnola Senese Marble, that costly stone is cleverly supplemented by scagliola, a plaster-based look-alike.

gorgeous Montagnola Senese Marble, the forest of columns and pilasters is faced instead in magnificently mimetic *scagliola*—a type of faux stone crafted on site by highly skilled Italian émigré artisans.

Scagliola (pronounced scal-YOH-lah) is a plaster-based substance that includes such ingredients as gypsum, pigments, isinglass, and alum. When blended together, these create a paste that can be set in molds or applied to a surface and then, when thoroughly hardened, polished to a high sheen. Various methods are used to simulate the stone being copied. For example, strands of silk dipped in coloring can be drawn through the mixture to duplicate the true marble's pattern of veins. However complicated and time-consuming the production of this stand-in for stone might seem, it is, all told, considerably less expensive than covering an entire interior with fancy imported marble. In essence it's a cost-saving measure, albeit a very high-class one found in some of the most magnificent buildings in Europe and America. We'll see it again in another guise at the Second Ward Savings Bank (site 6.17).

The rotunda's genuine Montagnola Senese Marble is best examined in the balustrades and stairway handrails. This gorgeous and highly sought-after metamorphic rock, often called "Giallo Siena," has been quarried since at least the Middle Ages in the low hills northwest of Siena, Italy—that incomparable medieval city set so sweetly on the mythic landscape of Tuscany. Originally a Jurassic limestone that formed on the floor of the Tethys Ocean, the Montagnola Senese was transformed into marble in the Eocene epoch by compressional forces that formed the Apennine Mountains. Its overall yellow tint is the result of the iron-oxide mineral goethite suffused through it. It also sports dark, squiggly stylolites and white calcite veins that add character and complexity to its patterning.

6.13   St. James Episcopal Church

833 W. Wisconsin Avenue
Current name: St. James 1868
Completed in 1868
Architect: Gordon William Lloyd (mistakenly identified as Edward
     Townsend Mix in some sources)
Geologic features: Wauwatosa Dolostone, Prairie du Chien Group
     Dolostone

Now serving as an "events venue," this handsome Gothic Revival ex-church retains its mien of time-honored sanctity with the help of its random-coursed, rock-faced Wauwatosa Dolostone. However, another much rarer carbonate rock is also present, according to this site's National Register of Historic Places listing. This presumably reliable document notes that the Wauwatosa exterior is "trimmed with dressed stone from the Bridgeport quarries."

The first association any geologist well versed in our Regional Silurian Dolostone varieties would make is that the trim, visible in the door and window surrounds and other decorative detail, was obtained from the large quarry in Chicago's South Side neighborhood of Bridgeport. After all, Milwaukee and other southeastern Wisconsin towns certainly do have buildings that feature stone from northeastern Illinois. But apparently that's not the case here. Instead, this second St. James stone selection came from another more distant Bridgeport: the hamlet that lies near the scenic locale where the Wisconsin River flows into the Mississippi. Just next door to the city of Prairie du Chien, it's a rural community that once boasted "celebrated" quarries, "the most extensive" in southwestern Wisconsin, which extracted a lovely buff-to-yellowish arenaceous dolostone most suitable for architectural use. So highly regarded was it in the 1860s that it was used extensively for the third State Capitol in Madison. That domed building was mostly destroyed by fire in 1904 and was ultimately replaced by the present, even grander edifice, which, incidentally, is a showplace for fancy rock types foreign and domestic.

At Bridgeport the Prairie du Chien Group Dolostone was taken from one or both of two Lower Ordovician formations, the Oneota and the Shakopee. (Oneota produced not in Wisconsin but across the Father of the Waters in Minnesota can be found in several Milwaukee buildings cited in this book.) The Prairie du Chien Group is a collection of marine sedimentary strata that formed near the end of the episode of global sea-level highstand known as the Sauk Sequence. In contrast, the Wauwatosa Dolostone and the rest of our area's Silurian bedrock were deposited during the following Tippecanoe Sequence.

6.14 Court of Honor

W. Wisconsin Avenue between N. 8th and N. 10th Streets
Geologic features: Somesville Granite, Salem Limestone, Barre Granodiorite, Isabella Anorthosite, unknown granitoid, Bronze, Concrete

This linear procession of public monuments and sculptures occupies two traffic islands set in the broadened Wisconsin Avenue between the Central Public Library and Interstate 43. The following descriptions start with the easternmost point of interest and continue westward.

Washington Monument (1885; Richard Henry Park, sculptor)

The bronze ensemble featuring George Washington standing on high and a woman and child positioned below him was produced by foundries in Florence, Italy. In recent decades, until their refurbishment in 2016 and 2017, all three figures had a green copper-carbonate/sulfate patina. Now their previous brown copper-oxide color, which represents an earlier stage of the natural weathering process, has been restored. The pedestal, set on concrete steps, is itself of considerable geologic interest because it's made of Maine's Somesville Granite, quarried near the community of Mount Desert, on the island of the same name that is now the largest portion of Acadia National Park. Originally mapped as Devonian, the Somesville has more recently had its age recalculated to about 424 Ma, which puts it in the Upper Silurian instead.

In the days of its production the Somesville was extensively quarried and marketed widely in the East, but it's one of the rarest granites found in Milwaukee. While it's pinkish when viewed close at hand, at a distance it's buff, an unusual tint for a granitoid, and quite reminiscent of Salem Limestone and Berea Sandstone. This offbeat coloration is due to its mineralogical makeup, which includes the cream to pink alkali feldspar orthoclase, gray plagioclase feldspar, black biotite, and glassy quartz. You'll quickly discover that some of the pedestal's biotite is segregated in larger clumps, which quarriers call *knots*. Often considered unsightly, they here add some interest to an otherwise uniform texture.

Spanish-American War Soldier (1932; Ferdinand Koenig, sculptor)

The medium- to coarse-grained igneous stone of this bronze statue's pedestal has not been identified. But it still deserves scrutiny: it contains large labradorescent crystals that glitter on sunny days, despite the current coating of soot. The statute itself, representing a rifle-toting, campaign-hatted infantryman, was made by Chicago's American Bronze Company. In contrast to the reconditioned Washington statue's brown coating, it still wears a brighter green copper-carbonate/sulfate patina.

Midsummer Carnival Shaft (1900; Alfred C. Clas, architect)

Freudian associations are hard to avoid when one realizes this solitary column was erected to honor an annual civic celebration presided over by an official designated Rex, the "Ruler of the Kingdom of Pleasure." This festival dedicated to secular delight must have proven too much for Milwaukeeans of that era, because the Midsummer Carnival only survived for a few years. Or, perhaps, it was just too hot and humid to get sufficiently into the swing of things. At any rate, in a city replete with so many other examples of "Bedford Stone"—here read Salem Limestone—this monument might seem nothing more than a geological also-ran. Actually, though, it's a superb place to see *crossbedding*, a striking sedimentary structure found in rocks composed of grains deposited by fluid in motion.

The crossbedding here can most readily be found in the basal portions of the pedestal, where you'll see parallel diagonal lines that were laid down by tidal-shoal currents in the Mississippian-subperiod sea covering southern Indiana approximately 340 Ma ago. The direction the water was flowing is indicated by the lines' downward slopes. And, while you're at it, take a moment to examine the multitude of tiny fossil fragments that, with their calcite matrix, make up this stone.

The Victorious Charge (1898; John S. Conway, sculptor)

This dramatic work of art, the most effective and moving of the lot, is an essay in the adrenalized desperation of battle. Like so many other Civil War monuments that date to the 1890s, it was commissioned by a generation of Americans who still remembered what a titanic, blood-soaked, and heart-rending experience that fratricidal war had been. The commemorators felt obligated to acknowledge and honor those swept up in the conflict. Contrast this respectful understanding with the modern travesty of ineptitude committed by a local Public Radio station, which posted an online article identifying the memorial as *Victoria's Charge* instead. Personally, I doubt if any of its three still-advancing soldiers or their fallen comrade would have answered to that name. But if one of them actually did, we must assume, given the prevailing Victorian mind-set of the time, that it was Victoria's secret.

While the bronze statue itself was the work of the Crescenzi Foundry of Rome, Italy—sculptor Conway resided in the Eternal City at the time—the pedestal is a product of Vermont, that small state with a gigantic tradition of architectural-stone extraction. The rock type used here is the Barre (pronounced "Berry") Granodiorite, a granitoid that has the majority of its feldspar content in the form of plagioclase. Commercially quarried since the early 1800s, it's undoubtedly one of America's most famous, widely used, and highly regarded monumental stones. Even in Wisconsin, one of the Green Mountain State's major competitors in this industry, the Barre can be found in various Civil War memorials, from Kenosha to Baraboo to this one. And it's difficult to locate any major graveyard in the state where the Barre isn't present in the form of headstones or mausolea, as we'll see in Milwaukee's own Forest Home Cemetery (site 7.5).

FIGURE 6.4. The most evocative and engaging monument in Wisconsin Avenue's Court of Honor, *The Victorious Charge* comprises a bronze quartet of Civil War soldiers set on a plinth of Barre Granodiorite. Behind it at left stand the spires of the Catholic Church of the Gesu (site 6.23).

The Barre Granodiorite is a fine- to medium-grained intrusive igneous rock that has a pale-gray, salt-and-pepper aspect due to its mixture of off-white plagioclase feldspar, gray quartz, and black biotite mica. Devonian in age, it owes its existence to the Acadian Orogeny, the mountain-building event triggered by the collision of Laurentia with the microcontinent Avalonia.

Flower Bed Edging

Easily overlooked but every bit as geologically significant as the Court's statues is the low stonework edging around the flower beds. Of simple style and presumably fairly recent installation, these low barriers are made of a porphyritic igneous rock with light-green phenocrysts, some quite labradorescent, swimming in a finer, dark-green matrix. While I've so far found no records that cite its identity, in this case I'm willing to guess, because it's a dead ringer for a handsome selection marketed as "Lake Superior Green"—in other words, the Isabella Anorthosite.

As I write this, I have before me a polished slab of this rock type kindly provided by Minnesota's Coldspring Corporation, the major stone producer that exclusively quarries it. And the specimen is a perfect match with the edging. If the latter turns out to be something else, it means there are two commercially available stone varieties that are completely identical to the human eye, at least this side of a petrologist's laboratory. But that's unlikely, so let's assume the edging stone is indeed the Isabella.

Quarried in the wilds of Lake County, Minnesota, the Isabella Anorthosite is part of the Duluth Complex, an extensive assemblage of mostly mafic igneous rock produced during the Midcontinent Rift episode described in chapter 2. This means it dates to the final chapter of the Mesoproterozoic era, about 1.1 Ga. Anorthosite is of tremendous interest because of its manifestly weird composition, the majority of which is just plagioclase feldspar. Geologists speculate that it forms as a *cumulate*, a mass of low-density crystals floating atop a body of still-molten basaltic magma, like the skin on boiled milk. Also notable as a major constituent of the Lunar Highlands, it's thought to have originated there at the surface of the nascent moon's lava ocean. But in the case of the Isabella variety, the anorthosite came into being during one of our own planet's most impressive rifting events, as an intrusion in a greater expanse of gabbro, diabase, and basalt.

6.15   Germania Building

135 W. Wells Street
Current name: Germania Apartments
Completed in 1896
Architectural firm: Schnetzky & Liebert
Geologic features: Wauwatosa Dolostone, Salem Limestone, American Terra Cotta, Copper

Considered a Beaux Arts design, albeit one with a decided Teutonic twist, the Germania Building certainly has more than its share of different building and ornamental materials on exterior display. Of these, the three that have not been conclusively identified are the pink, coarse-grained granitoid of the damp course, the finer, very light gray granitoid of the entrance columns, and the pale-tan pressed brick of the third story and above. Fortunately, all else can be sourced, though there has been some confusion in the local literature about what exactly goes where.

Constituting only the basement walls, and visible only in the northern-elevation stairwell, is the locally quarried Wauwatosa Dolostone. In a much more visible zone above the damp course on the first and second floors, the rock type is not more Wauwatosa, as one reference suggests, but the ever-popular Salem ("Indiana," "Bedford") Limestone. At the time of the Germania's construction, the abundantly available, relatively inexpensive, and remarkably workable Salem was already often outcompeting the local dolostones of Milwaukee- and Chicago-area producers. Unlike the granitoid directly below it here, it cannot take a polish, but its ashlar has nonetheless been fancily rusticated and dressed in both smooth-sawn and comb-chiseled forms.

Farther up, you'll also spot ornament and trim fabricated by the American Terra Cotta Company. This firm, located in what is now Crystal Lake, Illinois, seriously challenged the regional dominance of Chicago's Northwestern works by producing terra-cotta of comparable quality, as discussed more fully in the Watts Building (site 5.21) section. And at the Germania's summit are its famous "Kaiser's helmet" cupolas, Milwaukee's most iconic architectural reference to the immense role played by the city's German American community. These are sheathed in copper, the remarkable red metal, mined from such ores as chalcopyrite and chalcocite, that's been in human use for at least 10 ka. Its once-gleaming metallic surfaces are now bright green after decades of interaction with the atmosphere. This distinctive color signals the fact that the surface of the original elemental copper has gone through the usual progression to pink to dark-brown copper oxide, and then gradually on to the present and final form of copper carbonate with possibly some copper sulfate as well.

6.16   Milwaukee Public Museum

800 W. Wells Street
Completed in 1962
Architectural firm: Eschweiler, Eschweiler & Sielaff
Geologic features: Salem Limestone; geology exhibits

As this book goes to press, the Milwaukee Public Museum (MPM) is planning to move to a new location at the northeast corner of McKinley and 6th Streets in late 2026. In the meantime, a visit to the present facility is a must for the urban naturalist. The building's exterior cladding, southern Indiana's Salem Limestone, is indeed a very familiar sight in Milwaukee County, but given its fossil content it's always worthy of close examination.

Still, the main attraction is indoors. The first-floor permanent exhibit titled *The Third Planet*, centered on the epic theme of plate tectonics, includes an utter masterpiece of educational artistry, the Silurian Reef Diorama. To stand in front of this magnificent re-creation of the local seafloor of 430 Ma is to gaze in won-der at this region's first architects: a diverse array of corals, whose building skills formed the framework of an ecosystem that also included brachiopods, ortho-conic nautiloids—the top predators here—trilobites, crinoids, and various other forms of ancient marine-invertebrate life. Other sections of the exhibit acquaint visitors with the Pleistocene Ice Age (don't miss the walk-in glacier) and other stops in the Carboniferous and Cretaceous periods.

Once the new museum opens its doors, it will feature a Time Travel Gallery that offers visitors a redesigned version of the Silurian Reef, as well as exhib-its on a Wisconsin Cambrian trilobite trackway, Mesozoic dinosaurs, and the Pleistocene-epoch Hebior Mammoth unearthed in Kenosha County.

And for those wishing to learn more about the ancient underpinnings of the Milwaukee and Chicago regions, the MPM also maintains *The Virtual Silurian Reef*, a website I highly recommend and list in the Selected Bibliography.

6.17 Second Ward Savings Bank

910 N. Dr. Martin Luther King Jr. Drive
Current name: Milwaukee County Historical Society
Completed in 1913
Architectural firm: Kirchoff & Rose
Geologic features: Salem Limestone, Carrara Marble, Scagliola, Lar-
    issa Ophicalcite

The pleasant and productive hours I've spent researching this book in the Histor-ical Society archives here have probably biased me beyond all hope of objectivity. But I think this marvelous edifice, like the similarly triangular City Hall, is one of Milwaukee's supreme geologic treasures.

At first its exterior may seem geologically unremarkable. Once again, the neo-classical elements of the Beaux Arts style have been expressed in the monolithic medium of Salem Limestone—the building trades' "Bedford Stone" or "Indiana Limestone." But I have yet to see one architectural exposure of this ubiquitous and seemingly bland Hoosier rock that does not reveal, even to the most jaun-diced gaze, some fascinating detail—telltale wisps of crossbedding, or fossils ranging from tiny forams to bits of crinoids, bryozoans, brachiopods, and other creatures of the Mississippian-subperiod sea. With the Salem, boredom is not an option. One must always approach it with both reverence and a hand lens.

That said, in this case the greatest discoveries lie within. A lingering visit to the interior of the Historical Society's magnificent headquarters is absolutely obligatory. The most obvious geologic point of interest you'll see on entering is the white and purple-veined wall panels, most noticeable flanking the clock above the massive vault. These are splendid examples of "Pavonazzo," one of the most sought-after varieties of Italy's Carrara Marble. (It's also rendered as "Paonazzo" and various historian-mangled versions thereof.) Unlike the Carrara's pure-white, waxy "Statuario" grade preferred by sculptors, the stone displayed here is highly brecciated. This is a good indicator that the marble's locale of origin, the Apuan Alps, has seen a high level of crust-rending tectonic activity since this rock, originally a Late Triassic to Early Jurassic limestone, was metamorphosed in the Oligocene and Miocene epochs.

The dark infillings of the "Pavonazzo" look like bolts of electricity jumping and crackling across the stone's surface, and the dramatic imagery they impart has been put to brilliant use in the clock surrounds. I've seen book-matched stone panels in buildings from here to Istanbul, but nowhere else have I beheld such a stunning result. This technique involves mounting sequentially cut stone sections next to each other, like paired pages of an opened tome, to produce a symmetrical pattern reminiscent of a Rorschach blot. But here we have not the

FIGURE 6.5.   The ancient art of book-matched ornamental-stone panels is stunningly displayed in Kilbourn Town's Second Ward Savings Bank. The heavily veined rock type used here is the "Pavonazzo" variety of Tuscany's Carrara Marble.

usual double-panel match, impressive enough in itself, but three adjoining sets of quadruples that create the illusion of receding rhomb-shaped tunnels. It's obvious that the bank's original owners were not averse to investing in exquisite stone and superbly skilled masons. Nor were they shy about subjecting their customers to a whopping dose of lithic shock-and-awe.

On the other hand, the designers did resort to one time-honored form of cost-cutting sleight-of-hand. The shafts of the interior's impressive Ionic columns seem to be made of another, fainter-veined marble, but as a matter of fact they're scagliola, the plaster substance that artisans skilled in its use can mix, tint, pattern, and polish into a simulacrum of stone that can fool even a seasoned petrologist, at least at first. This stand-in for fancy marble is more fully described in the Milwaukee Central Public Library section (site 6.12).

However, there's one more genuine stone type to be found in the column bases and some of the walls' base coursing. And it's another breccia, albeit one much darker in aspect. While its name was not recorded here, it's almost surely the Larissa Ophicalcite, a famous Hellenic stone much favored by Byzantine architects in such masterpieces as Constantinople's Hagia Sophia. Its identity can be determined by its white marble clasts, many of which have a mint-green coating of the mineral serpentine. They float here and there between much darker chunks of serpentinite, an exotic metamorphic rock derived from an ultramafic igneous type called dunite. This was chemically altered by contact with water, either at a midocean ridge or in a forearc subduction zone, and then ultimately plastered onto a continental margin—in this case, the amazingly complex crustal jigsaw puzzle that is the Aegean-facing underside of Europe.

The Larissa Ophicalcite, quarried near the major city of that name in northern Greece, is Jurassic or Cretaceous in age. Known to ancient Roman architects as *lapis atracius* and to their successors in the Eastern empire as the Greek equivalent of "Green Thessalian Stone," it's now most frequently marketed as "Verde Antico," a legacy of its popularity with Italian builders in the centuries since. While the Larissa was quarried in modern times, its production has apparently ceased in recent decades.

6.18 Milwaukee County Courthouse

901 N. 9th Street
Completed in 1931
Architect: Albert Randolph Ross
Geologic features: Salem Limestone, Tivoli Travertine

One of Wisconsin's most famous sons, Frank Lloyd Wright, dubbed this great looming beast of a building "the million dollar rockpile." But its rock is much too imposingly arranged and carved and becolumned to fit that description. What it really resembles instead is one of those hypertrophied designs Nazi architect Albert Speer cranked out for Hitler's proposed rebuilding of Third-Reich Berlin. The Courthouse, a lost-in-time Beaux Arts holdover, somehow materialized on its high perch smack-dab in the age of Art Moderne. A stylistic dinosaur, perhaps, but one that definitely roars out "Look on my Works, ye Mighty, and despair!" to the mere mortals scurrying like ants below it. The Badger State is blessed with many county courthouses of fine design that exude a democratic air of accessible officials and due legal process. But this one, whatever the civic attributes of the judges and clerks within, does not. How exactly a metropolis as architecturally accomplished as Milwaukee ended up with this particular object is a thing of wonder.

Suffice it to say that its exterior is 36,000 blocks of Salem Limestone, mortared into a much less congenial form than that of the Second Ward Savings Bank. The first-floor interior is worth a quick look, though, because of the Tivoli Travertine found in its floors and some of the wall paneling. For more on this spring-formed carbonate sedimentary rock see the US Bank Center (site 5.1) description.

6.19   Milwaukee Journal Sentinel Building

333 W. State Street
Previously the Milwaukee Journal Building or Journal Company Build-
    ing; now a Milwaukee Area Technical College student residence hall
Completed in 1924
Architect: Frank D. Chase
Geologic features: Oneota Dolostone, Cast Iron

No town east of the Mississippi beats Milwaukee as a showplace for Minnesota's Oneota Dolostone, in both its "Kasota Stone" and "Winona Travertine" varieties. This rock type's chief ornamental attributes are its unusual coloration, which ranges from buff to pale pink to pinkish yellow, and interesting textures that include vuggy pitting reminiscent of true travertine and, sometimes, intricate trace-fossil branching patterns. Here at the Journal Sentinel Building, both the pitting and the pinkness are apparent, as is the Oneota's excellence as a carving stone. The intricate detail of the high-arched entrance and the decorative bas reliefs of the façades represent some of the city's finest examples of sculpture in stone. This marine chemical sedimentary rock, discussed in greater detail in the descriptions of the Marcus Center (site 5.33) and Eagles Club (site 6.25), is Ordovician in age and outcrops from north-central Illinois to southwestern Wisconsin and its main quarrying region south of Minneapolis.

While the granitoid rock that constitutes the thin, easily overlooked grade course under the Oneota was not documented, there is one other, much more conspicuous geologically derived material displayed here. We're supposed to think that the elegant entrance lamps, doorway trim, and spandrels are that costly, classy alloy, bronze. After all, the metal seems to have weathered to that same lovely shade of copper-carbonate green seen on some of the city's most venerable buildings and outdoor statues. In fact, though, this is cast iron cleverly painted to simulate it. This sort of mimicry of materials, seen again and again in architecture, might be beauty on a budget, but it produces no less impact on the human eye.

6.20   Turner Hall

1034 N. Vel R. Phillips Avenue
Alternative name: Turnverein Milwaukee
Completed in 1883
Architect: Henry C. Koch
Geologic features: Cream City Brick, Wauwatosa Dolostone

FIGURE 6.6.   The Turner Hall is a splendid example of the Classic Milwaukee Formula version that couples Wauwatosa Dolostone and Cream City Brick. And in this case there's an added ingredient—nicely contrasting red roofing and trim.

A notable Richardsonian Romanesque design, this headquarters for Milwaukee's chapter of a nationally prominent German American athletic, political, and social society is one of the city's most *gemütlich* examples of the Third Version of the Classic Milwaukee Formula. Its cheerful, only moderately stained Cream City Brick is grounded with rock-faced ashlar blocks of what appears to be the locally quarried Silurian Wauwatosa Dolostone. The red brick used for trim, a laudable addition to the other two materials, has not been sourced.

6.21 Trinity Evangelical Lutheran Church

1046 N. 9th Street
Completed in 1878
Architect: Frederick Velguth
Geologic features: Cream City Brick, Wauwatosa Dolostone

This Victorian Gothic house of worship suffered a major fire in May 2018 that destroyed its nave roof and much of its beautiful interior. The good news for all lovers of this city's top-tier architecture is that the restoration process is, at time of writing, well under way. And the church remains Kilbourn Town's best demonstration of Cream City Brick that is very thoroughly darkened by decades' worth of accumulated soot. As noted in the Old St. Mary's Church (site 5.27) description, I think this adds character and conveys a noble acceptance of transformation wrought by the passage of time. Tidier minds than mine are bound to disagree. For them the only valid state is the original.

In a sense, Trinity Evangelical fits the Third Version of my Classic Milwaukee Formula because its design also includes what is apparently Wauwatosa Dolostone, though all sources so far consulted merely list it generically as "limestone." But unlike Turner Hall discussed in the preceding section, this locally produced carbonate rock is not a significant part of the church's appearance; it can be seen only where the building's basal walls are exposed because of the downward slope on the eastern side.

In addition, the National Register of Historic Places nomination form for this site alleges that the exterior trim, which is buff in color, is "Illinois sandstone." But architectural historians so frequently confuse sandstone with limestone and dolostone that this reference is hardly reliable. It is possible, however, that the rock type described is actually Lemont-Joliet Dolostone quarried in the Lower Des Plaines River Valley southwest of Chicago.

6.22   Pabst Brewery Complex

N. 9th Street and W. Highland Avenue and vicinity
Completed in different sections between 1858 and 1933
Architect: Otto Strack
Geologic features: Cream City Brick, Wauwatosa Dolostone

What has recently been repurposed for a variety of residential and business uses was at one time the largest beer-making operation on Earth. In strolling through the complex you'll understand more fully than anywhere else in town the sheer immensity of the Cream City Brick that was produced just for Milwaukee's brewing industry. Some buildings here also have plinths of rock-faced Wauwatosa Dolostone. On clear winter days, and in other times of low sun angle, these now-quiet blocks of buildings give off a sort of mythic, end-of-history glow.

6.23   Catholic Church of the Gesu

1145 W. Wisconsin Avenue
Original name: St. Gall Catholic Church
Completed in 1893; entrance portico added in 1902
Architects: Henry C. Koch & Company (1893); Herman J. Esser (1902)
Geologic features: Salem Limestone, Athelstane Granite

In a city boasting many handsome churches of Gothic Revival style, this is the one that in my mind stirs the deepest memories of its European predecessors. One can easily imagine it standing not on a major thoroughfare transecting a modern American university campus, but rather on a raised base facing a market square hemmed in by medieval flats, cafés, shops, and mazelike alleys. Its impressive but light-toned façade is Milwaukee's finest ecclesiastical use of the Salem ("Bedford," "Indiana") Limestone as dressed- and rock-faced ashlar and intricately carved ornament.

However, the church's geologic high point is the group of arch-supporting entranceway columns made of Athelstane Granite, of Paleoproterozoic age. This is the same handsome coarse-grained variety, quarried in Athelstane, part of northern Wisconsin's ancient Pembine-Wausau Terrane, that constitutes the lower portion of the Federal Building a mile to the east. But here the column shafts have been polished to fully reveal their assemblage of hefty feldspars—pinkish-buff microcline and a white plagioclase—mixed with black biotite mica and translucent quartz. Rarely has this planet's magma crystallized into so fine a form.

6.24   Tekton Series Sculpture

1234 W. Tory Hill Street
Completed in 1977
Sculptor: Ernest Shaw
Geologic features: Murphy Marble, Weathering Steel

You'll find this modestly sized abstract composition on the Marquette University campus, in a parklet just west of Eckstein Hall and north of the Haggerty Museum of Art. The work gives the urban explorer an intimate look at more than four decades' worth of weathering-steel oxidation. Here the iron-carbon alloy that has played such a titanic role in human civilization generously shares its plentiful crop of rust with the rock slab attached to it. And that rock in turn is one of America's great building-stone varieties—one that for some reason is rather underrepresented in Milwaukee's built landscape. In Chicago, it's been used as a primary design element for Buckingham Fountain, a church located in an urban canyon of the Loop, and even a Richardsonian Romanesque mansion façade. But here, it seems, architects and officials have decided Cream City monuments and buildings just don't look that pretty in pink.

And decidedly pink it is, with generous streaks of silver that signal its significant graphite content. Specifically, it's the "Etowah" brand of Georgia's famous Murphy Marble. Quarried in Pickens County, in the Peach State's portion of the Piedmont physiographic province, the Murphy is also offered in white-veined and unveined forms. Regardless of its exact coloration and patterning, it can generally be distinguished from its competitors produced in Carrara, Vermont, Alabama, and Colorado by its coarser texture, which is composed of macroscopic calcite crystals. If you look closely here, you'll see those interlocking grains are indeed distinctly visible.

This nonfoliated metamorphic rock began, probably in the Cambrian or succeeding Ordovician period, as a marine-limestone protolith mantling the Laurentian continental shelf that fronted the now-vanished Iapetus Ocean. Later in the Paleozoic the limestone was transformed into its current identity by pulses of crustal compression caused by the three major episodes of mountain-building that affected what is now eastern North America.

6.25   Eagles Club

2401 W. Wisconsin Avenue
Current name: The Rave / Eagles Club
Completed in 1927
Architect: Russell Barr Williamson
Geologic features: Oneota Dolostone, Salem Limestone

This Mediterranean Revival building, described by its current management as "a live-entertainment complex," was originally the home of the Milwaukee branch of the Fraternal Order of Eagles. Chromatically speaking, it's an absolute screamer, the best advertisement for the "Kasota" variety of Oneota Dolostone I've seen.

From a distance, its lovely and attention-grabbing carbonate rock, quarried in southern Minnesota, ranges in color from pale yellow and peach to burnt orange. As one approaches, though, it also reveals an unusual pockmarked texture that at very close range resolves itself into a fascinating array of branched tubes—a feature found in various marine sedimentary strata. This intriguing pattern was formerly described as *fucoidal* because it resembled some forms of marine algae. But when I checked with Phanerozoic-rock expert Andrew Retzler of the Minnesota Geological Survey specifically about the stone on display here, I learned that nowadays this feature is considered a trace fossil of the genus *Thalassinoides*. As such, it's thought to be a network of intricate, interconnected burrows made by benthic creatures when this rock was still loose sediment at the bottom of the epeiric sea that covered part of the Midwest in the Lower Ordovician period.

The Eagles Club also has its share of that great multitasker of architectural rock types, Indiana's Mississippian-age Salem Limestone. While it's about 140 Ma

FIGURE 6.7. The Oneota Dolostone of the Eagles Club exterior owes its puckered, branching textures to the presence of fossilized burrows of marine animals that lived half a billion years ago.

younger than the Oneota, it too formed at a time of high global sea level when our continent's midsection was covered in saltwater. Here it serves not as the exterior's main cladding but as nicely contrasting and exquisitely carved panels and window ornament, and a rooftop loggia as well.

6.26   Sheridan Apartments

2435 W. Wisconsin Avenue
Completed in 1927
Architectural firm: Backes & Uthes
Geologic features: American Terra Cotta, Copper

The Mediterranean Revival style seen next door at the Eagles Club continues here, though the building materials that express it are distinctly different. The most notable aspect of the Sheridan façade is the riot of terra-cotta cladding, ornament, and quoins fabricated by the American works. This company, situated in the rolling, morainal landscape of rural McHenry County in northeastern Illinois, derived its base clay from Lemont Formation till laid down by the Wisconsin ice sheet near the end of the Pleistocene epoch. This firm developed the Pulsichrome method of finishing its glazed cladding units using a pneumatic sprayer that could apply three colors at once. Various mottled patterns, including the one you see here, could be produced. Some of them closely mimicked sought-after varieties of dimension stone.

The source of the Sheridan's buff-and-browner brick is unknown, but one other geologic material of note, copper, is present on the building's cupolas and roof trim. It has weathered to the characteristic green carbonate/sulfate patina.

6.27   Central United Methodist Church

639 N. 25th Street
Completed in 1980
Architect: William Wenzler
Geologic features: Concrete, Fieldstone

In Milwaukee as in other cities, many buildings succeed in being both beautiful and geologically significant. Here we can only address the latter attribute. That's not to say this Brutalist church isn't architecturally significant, too. Its design, laudably guided by the principle of energy efficiency before that concept became a major trend, incorporates such now-common features as solar-power collection and a green roof.

Concrete can be a remarkably attractive material—one has only to think of that miracle of grace and transcendent geometry, the Quadracci Pavilion (site 5.40). But here the simple addition of vertical scoring to the exterior's sections has done little to convince the onlooker that this structure is a sacred space rather than a partially disguised command bunker or the surface expression of an ICBM silo. And one could plunk it down on the set of *Blade Runner 2049* without a second thought.

However, no urban geologist can or should resist spending a few minutes with the rounded to subangular fieldstone boulders that armor the church's Wisconsin Avenue slopes. These large chunks of rock, mostly either felsic or mafic igneous types, are *erratics*, detached larger pieces of stone hefted from their outcrop locales by the Pleistocene ice sheet and carried southward, sometimes hundreds of miles, before nature's most powerful long-distance hauler melted away and deposited them in this region. Building and landscaping contractors have harvested these boulders in impressive quantities from farm fields and excavation waste piles in the glaciated portion of the state. They're a perennial architectural favorite: used in the nineteenth century for house construction and ornament; in the twentieth century and later as suburban design accents adorning driveways, plantings, and ersatz waterfalls on corporate "campuses."

6.28 Tripoli Temple Shrine

3000 W. Wisconsin Avenue
Completed in 1928
Architectural firm: Clas, Shepherd & Clas
Geologic features: St. Cloud Area Granite, Salem Limestone, Rookwood Ceramic Tile

One absolute requirement for any center of world-class architecture is a certain quantum of incompatibility. A dose of cognitive dissonance or even downright goofiness can be a welcome thing in a profession all too prone to pomposity, preciousness, and bean-counting commercial intent. And also, because amid our skyscrapers and strip malls we stand at the geographic, climatic, and cultural antipodes of Old World Islamic architecture, it's all the more essential we get some of that, and use it exactly where it seems to have no context whatsoever.

Here, thanks to the Shriners organization, that wonderful incompatibility, that utter lack of context, has been gloriously achieved. Behold Sinbad's Baghdad amid the beer and brats, the abstract artistry of sun-drenched lands amid the sleet and snowplows. This fantasy-ridden building is a wonder and delight, standing where it does.

The origin of the Tripoli Temple's brown and ocher bricks has not been identified, but it's safe to say the Shrine is composed of quintessentially American geologic materials that produce a Cream City riff on Iranian and North African themes. Starting with the wide entrance stairway, you'll discover that a new variety of the polymorphic St. Cloud Area Granite has been used for the treads and risers. This is its "Sauk Rapids Pink" brand, quarried in the eponymous town where the Mississippi River, still 60 miles upstream from Minneapolis, flows across the resistant roof of the East-Central Minnesota Batholith. This Paleoproterozoic intrusive rock is medium-grained and composed of a mixture of pale-pink to buff feldspars, clear gray quartz, and black hornblende and biotite mica that is, for some reason I can't quite put my finger on, especially attractive. In the 1920s and 1930s it was a popular choice for building steps, though I've also seen it used as damp coursing on a bank in Waukegan, Illinois.

Flanking the stairway and also forming contrasting trim to the magnificent tilework of the entrance is what by now we must consider an old and faithful friend, the Salem Limestone, in its most flamboyant setting yet. Make sure you carefully inspect the pair of camels resting contentedly at the base of each wall (see figure 3.3). They too are made from this most sculptable of rock types.

FIGURE 6.8. When it comes to Milwaukee's great non sequitur of architecture, the Tripoli Temple Shrine, resistance is useless. Impossible not to love, it's decked out in everything from Salem Limestone and Minnesota granite to Rookwood ceramics, and provides a dearly needed dose of noncontextuality.

These representations of modern-day, desert-adapted ungulates have two very good reasons, paleontologically speaking, for being here. For one thing, their direct ancestors first entered the evolutionary record in North America, and they remained restricted to this continent for about 40 Ma, from the Eocene to the Miocene, before one line managed to migrate to Eurasia. (No doubt paleo-camels roamed the wilds of Paleo-Milwaukee long before our species' arrival.) And for the other thing, these figures, being Salem Limestone, are made of the remains of Mississippian-age marine creatures that thrived in the diverse saltwater ecosystem of southern Indiana 340 Ma ago. These days both statues show some signs of weathering, but that ultimately destructive process has for the time being helped to expose some excellent fossil specimens, including a nice brachiopod valve over half an inch wide. Please do not touch or try to remove any of these. Besides being the property of the Shriners, they're an educational resource all visitors should enjoy and respect in hands-off mode. If you insist on getting a memento, pull out your camera or phone and take a photo instead.

While the Salem and St. Cloud make the Shrine a worthy venue all by themselves, it's the ceramic tile that does the most to produce the building's Arabian Nights persona by cladding the central façade and the onion domes in a glorious play of color. And the intricate, repeating aniconic patterns, one hallmark of Muslim architectural ornament, are mesmerizing. The tile, which replicates the look of fired-and-glazed-clay decoration produced for millennia in the Middle East, was actually fabricated in that fabled den of Buckeye exoticism, Cincinnati, Ohio. There the Rookwood Company, most famous for its highly sought-after pottery, also produced architectural ornament for such landmarks as the Tripoli Temple and Chicago's Monroe Building. When I contacted Rookwood official Jim Hibben, I learned that the source and geologic underpinnings of the clay used in both sites' tiles are a bit hard to nail down. As it turns out, that firm has obtained its base materials over the decades not only locally, from the Ohio River Valley, but also from Tennessee, Kentucky, and other eastern states. In all likelihood these were either *fluvial* (stream-transported) or *residual clays* that formed more or less in place from the weathering of rock or chemical alteration of soils, and which contained the suitable mixture of feldspar, quartz, iron, and kaolin to meet Rookwood's specifications. In contrast, the great terra-cotta works of the Upper Midwest mostly depended on either glacial deposits or Pennsylvanian-age shales and paleosols shipped in from the Illinois Basin. The fact that all these different geologic sources have been available to feed the architect's desire for beautiful building materials reminds us that we inhabit a hyperactive planet that tries its best to grind down everything and everyone to clay.

6.29   George Weinhagen House

3306 W. Highland Boulevard
Completed in 1911
Architect: Herman W. Buemming
Geologic feature: Cream City Brick, Roman Brick, Terra-Cotta Roof Tile

Sometimes the same building material can be classified in more than one way. This is especially true at the Weinhagen House. For here Cream City Brick is also Roman Brick, a type much favored by Frank Lloyd Wright and his adherents. Longer and lower than standard American brick, the Roman variety draws its name from a similarly shaped type employed by ancient Imperial architects. In its modern form, it's most usually a low-slung and elegant 3 5/8 x 1 5/8 x 11 5/8 inches. This sleek shape emphasizes the horizontality favored by Prairie School architects.

Another striking feature of this residence is its beautiful green-glazed terra-cotta roof tile. Surviving documentation does not indicate whether the tiles are original or not, or whether they were made by the most preeminent manufacturer, the justly renowned Ludowici Company.

6.30   St. Michael Catholic Church

1445 N. 24th Street
Completed in 1892
Architectural firm: Schnetzky & Liebert
Geologic feature: Wauwatosa Dolostone

My favorite among the city's Gothic Revival churches ever since I first gazed upward at its imposing verticality, high-steepled St. Michael's convincingly demonstrates how perfectly suited to this high-medieval style Silurian-period Wauwatosa Dolostone truly is. The rock-faced ashlar shows the mellow-yellow, character-building tints of well-weathered Racine Formation rock. This is the result of the Wauwatosa's original ferrous-iron impurities being oxidized and hydrated on prolonged exposure to the elements.

6.31   Old Main, Milwaukee Soldiers' Home

515 N. General Mitchell Boulevard
Original name: Main Building or Building 2, Northwestern Branch,
   National Home for Disabled Volunteer Soldiers

Completed in 1869; addition in 1876
Architect: Edward Townsend Mix
Geologic features: Wauwatosa Dolostone, Cream City Brick, New York
     Red Slate, Vermont Unfading Green Slate

This splendid Victorian Gothic structure, recently restored, occupies the most architecturally effective high ground in the city. Unlike the bloated authoritarian bulk of the County Courthouse, it does not use its exalted position to overwhelm and squash the soul of the onlooker. This book's most venerable example of the Classic Milwaukee Formula's Third Version, Old Main combines a low plinth of Wauwatosa Dolostone (not specifically identified as such in the literature, but extremely likely) to support Cream City Brick elevations. The Wauwatosa is laid in rough-faced, random-course fashion.

Taken as a whole, the mansard roof faces are a masterpiece of slate-tile artistry in both rectangular-overlap and fish-scale patterns. Once again, historical and restorers' descriptions do not name the exact varieties used, but in all likelihood two of the four or five types present can be identified as New York Red Slate and Vermont Unfading Green Slate. The former is the only red American selection in widespread use; and despite its name the latter often has the light-blue cast

FIGURE 6.9. This section of Old Main's upper elevation supplies us with a strikingly patterned example of the slate-roofer's art. Each color represented here—red, light blue, and darkest gray—indicates the varying geochemistry of the marine muds from which this rock type formed.

seen here. The remaining uncertain shades could be Vermont Purple and one of Vermont's gray or black brands, but other possibilities quarried in Quebec, Pennsylvania, and other Eastern locales can't be eliminated without good provenance clearing up the uncertainty. Both the Vermont Unfading Green and New York Red, and possibly one or more of the others, were quarried in the same area straddling the boundary of the two named states. For more on the origin and composition of the slates from this region see the Wisconsin Consistory (site 5.23) description.

6.32   Soldiers' Home Reef

Between Chrysler Wood Avenue and Frederick Miller Way; just west of
   American Family Field
Architectural firm: Silurian Corals, Sponges, & Associates
Geologic features: Ancient Reef, Racine Formation Dolostone exposure
Note: Please do not collect fossils or damage rock outcrops

In a work on architecture, why bother to mention this low wooded hill where no building stands?—Because the hill itself is a work of architecture, a profession that predates our species by at least half a billion years.

This is one of a famous pair of Silurian fossil reefs exposed in Milwaukee County's Menomonee River Valley. Both are now rightfully listed on the National Register of Historic Places. The Soldiers' Home site, just a three-minute walk from Old Main, is significant not for having a highly visible internal structure or accessible fossil contents, but rather for its overall structure. Reef rock, generally harder than surrounding interreef deposits, is more resistant to erosion and therefore tends to form mound-shaped topographic highs like this one. In the technical literature, such landforms are referred to as *reef-controlled hills*.

However, when the trees are not in leaf you can get a peek at the weathered Racine Formation bedrock on the reef's north flank, facing Frederick Miller Way. This is the same unit that provided human architects with the beautiful Wauwatosa Dolostone for some of this city's most distinguished buildings. Be very mindful of fast-moving auto traffic and stadium-lot parking restrictions, and do not stand in the roadway as you take in the view. This section of the reef is best approached via the paved footpath from public-access parking in the vicinity of Old Main.

While we can primarily credit the coral organisms and stromatoporoid sponges of 425 Ma ago for the construction of this intriguing landform, it's important to keep in mind that they were joined by a diverse community of

FIGURE 6.10.   Now a small tree-covered hill next to the Brewers' ballpark, the Soldiers' Home Reef is one of two Menomonee Valley sites that reveal the amazing architectural skills of corals and other Silurian-period marine organisms.

other marine benthic and nektonic animals so beautifully depicted in the great Milwaukee Public Museum reef diorama (see site 6.16). Among this cast of characters were nautiloid cephalopods, "sea lily" crinoids, two-valved brachiopods, "moss animal" bryozoans, and trilobites. These creatures inhabited a shallow, sunlit, subtropical sea on the western edge of the developing Michigan Basin; they built their underwater home when this portion of our continent was about thirty degrees south of the equator.

6.33   Schoonmaker Reef

Just north of 1225 N. 62nd Street, Wauwatosa
Architectural firm: Silurian Corals, Sponges, & Associates
Geologic features: Ancient Reef, Racine Formation Dolostone exposure
Note: Please do not collect fossils; no access to private property or
    within the protective fence without permission (see below)

This second National Landmark fossil reef in the Menomonee River Valley is located in what was formerly Wauwatosa's great stone-producing district. It features a long, quarried cliff face of Racine Formation dolostone that gives a much better idea of a reef-controlled hill's inner composition. That said, it can present the urban geologist who wants to see it up close with a daunting challenge. This is because it stands just north of an apartment compound, unsurprisingly called the Reef, that is private property. A bit of the cliff can be spotted from the north-eastern corner of the commercial strip-mall parking lot just to its southeast, but even so, intervening apartment grounds and a security fence may give the once-hopeful visitor that sinking, so-near-and-yet-so-far feeling.

In an attempt to figure out why this world-renowned paleontological site is so inaccessible, I contacted Wauwatosa city officials, who confirmed that no regu-larly scheduled tours are currently offered. However, I was also told that access within the fence is periodically granted to educational groups and interested individuals who enter at a gate at the eastern end. To request permission, contact Wauwatosa's Parks and Forestry Division at tparks@wauwatosa.net, or call (414) 471–8422. Alternatively, without trespassing you can catch a pretty good glimpse of one part of the reef exposure between buildings from the 62nd Street sidewalk, near the western end of the apartment complex.

As the region's preeminent Silurian experts Joanne Kluessendorf and Don-ald Mikulic noted in a detailed account of this site (see Selected Bibliography), Schoonmaker Reef has been explored, collected from, and described by an impressive array of scientists. This honor roll includes Milwaukee's own Increase Lapham, the great nineteenth-century paleontologist James Hall, the equally great Wisconsin geologist Thomas Crowder Chamberlain, Cedarburg-born Amadeus Grabau, and noted twentieth-century paleontology-textbook author Robert Shrock. Their work, and that of so many others, led to our modern under-standing of the appearance and greater geologic significance of these fascinating natural structures that were built in a very different version of Milwaukee County some 425 Ma ago. And the resulting inventory of identified fossil types, which includes bivalves, brachiopods, bryozoans, cephalopods, corals, gastropods, stro-matoporoids, and trilobites, has revealed the remarkable diversity of this ancient marine community. The architectural geologist keeps all that in mind but also realizes that nowadays Schoonmaker Reef is one of the few places one can still see the Wauwatosa Dolostone, featured in so many of Milwaukee County's finest buildings, as it appears in outcrop.

6.34 Fiebing House

7707 Stickney Avenue, Wauwatosa
Completed in 1925
Architectural firm: Arnold F. Meyer & Company

Geologic features: Crossville Sandstone, Lake Border Moraine swell-
and-swale topography

This and Milwaukee's Hoelz House (site 8.26) are two outstanding examples of
architect Meyer's use of the varicolored Crossville Sandstone as exterior cladding for modestly scaled homes. A rock selection both attractive and enduring,
the Crossville is a clastic sedimentary type quarried on Tennessee's Cumberland
Plateau. It formed in the Pennsylvanian subperiod from sand deposits eroded
from the mighty Alleghenian Mountains that then loomed just to the east. This
range, very possibly as lofty as the modern Himalayas, rose when northwestern
Africa collided with eastern North America during the assembly of the supercontinent Pangaea.

The Crossville Sandstone is best known in the building trades as "Crab
Orchard Stone" and "Tennessee Quartzite." While many commercial monikers
misrepresent their stones' true identities, that's not necessarily the case here.
This rock is indeed sometimes classified as an *orthoquartzite*. In other words,
it's a quartzite, but an unmetamorphosed one, composed of grains of very hard
quartz. These are firmly cemented together not with calcite or some other reactive mineral, but with silica—that is, with still more quartz. The Crossville is also

FIGURE 6.11.   Wauwatosa's Fiebing House, set on the rolling terrain of the
Lake Border Moraine. Its unusual flagstone cladding is made of Crossville
Sandstone from the Tennessee portion of the Cumberland Plateau.

renowned for its *Liesegang rings*, concentric patterns of darker lines and loops caused by the migration of iron-rich groundwater solutions. If you take a good look at the elevation of the projecting wing closest to the sidewalk, you'll see some hints of this feature. And note how the uncoursed rubble-style stonework imparts a rustic, cottagey look to this residence.

Also always of geologic interest is the greater context. Here, that's the gently sloping land on which the Fiebing House and its neighbors sit. This is a good example of the *swell-and-swale terrain* characteristic of Milwaukee County's Lake Border Morainic System, a series of broad, irregular ridges of till dumped at the margin of the Wisconsin ice sheet in the waning days of the Pleistocene epoch. The upper surfaces of end moraines such as this often form hummocks and depressions as the unsorted glacial sediments settle unevenly.

MILWAUKEE: WALKER'S POINT, BURNHAM PARK, LINCOLN VILLAGE, FOREST HOME HILLS, AND LYONS PARK

7.1   Our Lady of Guadalupe Catholic Church

605 South 4th Street
Original name: Holy Trinity Catholic Church
Completed in 1850; steeple erected in 1862
Architects: Victor Schulte (1850); Leonard Schmidtner (1862)
Geologic feature: Cream City Brick

For your own exploration of this venerable Zopfstil church, I wish you conditions and impressions of the kind I experienced on my initial visit here.

It was a late-winter weekday. At first I heard the loud clanks and clangs of machinery down the street, by no means out of place in this industrial neighborhood. And then, at the stroke of noon, Our Lady's bells rang out in the steeple high above: great waves of primeval sound pulsing through the chill lake haze. That and the sight of this soot-stained exposure of Cream City Brick finally closed the loop in my mind between the human, the divine, and the natural, with nothing less sublime before me than I'd seen in the wilds. In fact this *was* the wilds, a great deposit of river clay that had, by the fire of the kiln and tectonic force of the human hand, lifted itself more than a little way heavenward.

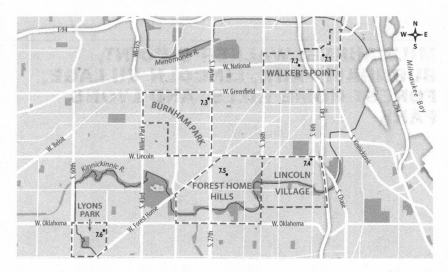

MAP 7.1. Sites in Milwaukee's Walker's Point, Burnham Park, Lincoln Village, Forest Home Hills, and Lyons Park neighborhoods.

7.2   John Burnham Block

907–911 W. National Avenue
Completed in 1875
Architect: Edward Townsend Mix
Geologic features: Cream City Brick, Cast Iron

It's twenty-five years younger than Our Lady of Guadalupe and was built with no sacred intent, but the Burnham Block shares with that venerable Walker's Point church a sense of grimy timelessness that only Cream City Brick, seasoned and speckled with fossil-fuel soot, can impart.

In fact, no Cream City Brick building has a greater right to be in this book. Its original owner, John Burnham, was part of the family that, in two separate branches and companies, ran brickmaking works just to the northwest of here. These great industrial operations exploited the massive deposits of fluviatile and glacial clays of the Menomonee River floodplain north of National Avenue— deposits that were, in some places, 100 feet thick. From them rose not only this multipurpose storefront–meeting hall, but many of Milwaukee's more illustrious structures as well.

Also worth a look here are the still-extant cast-iron columns of the Burnham Block's ground-floor façade. They're a reminder that this melted-and-molded form of the world's most abundantly mined metal was a favorite ornamental and structural material in nineteenth-century American architecture.

7.3   St. Joseph's Convent Chapel

1501 S. Layton Boulevard
Completed in 1917
Architectural firm: Brust & Philipp
Geologic features: Salem Limestone, Flemish Bond Brick, Rock-Faced Brick

FIGURE 7.1.   The Layton Boulevard elevation of St. Joseph's Convent Chapel boasts a harmonious blend of intricately carved Salem Limestone and Rock-Faced, Flemish Bond Brick. Note the two lippos, described in the accompanying text, that watchfully guard the arched entrance.

While this site's gorgeous interior purportedly boasts fifteen marble varieties—"marble" in the looser, architectural sense—the exact identity of only one, the Carrara of the three altars, seems to have been recorded in the literature. This and the fact the chapel is not routinely open to the public makes it best to concentrate on the façade instead. It's Romanesque, but not Richardsonian; it emulates medieval European church models without incorporating the idiosyncrasies of the great American architect's designs.

Though the main entrance is actually within the convent, the arched doorway facing Layton Boulevard, and indeed the entire ground level, have by no means been neglected. They are smartly decked out in pale-buff Salem ("Bedford," "Indiana") Limestone. The ashlar is dressed-faced; the intricate ornament above the door is a splendid essay in ecclesiastical zoology and botany. The repeating acanthus-leaf motifs of the arch itself are surrounded by a bestiary that includes grazing, be-antlered stags, the Lamb of God, and my favorites, a pair of *lippos* (what I take to be lion-hippo hybrids). Could any rock variety besides the Salem give rise to such a wonderful ecosystem in stone?

The ocher brick of the front's higher portions has not been sourced, but it is nonetheless an instructive example of Flemish Bond work, where masons have created a simple but striking pattern of alternating stretchers with headers. Generally, the headers of one course are centered on the middle of the stretchers of the courses directly above and below them. And another subtle effect is present, too. Note how the bricks are not really flat-faced, but rather have rough, slightly projecting surfaces like quarry-faced stone. This use of Rock-Faced Brick imparts an interesting texture missing from more perfunctorily designed buildings.

7.4   St. Josaphat Basilica

2333 S. 6th Street
Original name: St. Josaphat Church
Completed in 1901
Architect: Erhard Brielmaier
Geologic features: Buena Vista Siltstone, "Maine Granite," Copper

It could be argued that a structure this utterly grand deserves a grander setting: perhaps atop one of eastern Wisconsin's moulin kames, where, like the Holy Hill shrine 25 miles to the northwest, it could lord over a leafy landscape of wider vistas. But even here and from the nearby interstate St. Josaphat is the most impressive structure for many blocks around. Consecrated in 1929 as Milwaukee's only Roman Catholic basilica, this house of worship is rightly extolled by various

FIGURE 7.2.   The St. Josaphat Basilica is both an imposing house of worship and an impressive monument to community pride, priestly perseverance, and canny architectural recycling. And it is, in terms of its exterior stone, the resurrection of two now-demolished Chicago buildings.

architectural histories and guides as an almost unbelievable grassroots success story, a model of magnificence achieved thriftily, thanks to founding pastor Wilhelm Grutza's canny acquisition of recycled building materials, and to outstanding financial support from the church's predominantly Polish immigrant community. Those materials were acquired not locally but in Chicago, and the haul included 200,000 tons of stone and various doors and fixtures, all shipped up by rail on a holy armada of 500 freight cars. According to one account, the bill for the lot was just $20,000—an astoundingly low figure, even in that day, for such an impressive construction-goods inventory.

Accounts of the basilica's origins correctly point out that those salvaged materials include the exterior's main sedimentary ashlar and six great crystalline-rock columns of the colossal portico. But there is considerable confusion about where some of these materials—specifically the columns—came from. Most narratives suggest or overtly state that both rock types were retrieved from the demolition of Chicago's ill-fated US Custom House, Court House, and Post Office.

This massive edifice, completed in 1880, was set on a shallow, continuous-raft concrete foundation soon undermined by the differential settling of wetland sediments every bit as treacherous as downtown Milwaukee's. Windy City journalists of the time delighted in recounting, in the florid style of the time, its long litany of woes. An 1893 Rand McNally guidebook probably put it best: "The dark Gothic mass which arose was too heavy for the soil, and sank steadily. It was bolted together, but continued to sink, amid the lamentations of office-holders who would not flee from the fate they feared. Courts have adjourned precipitously on the loud report of an opening to the walls, or the flooding of a water-pipe, and the tiles in the floors respond with a melancholy rattle as the citizen hurries through the corridors to escape the Post Office draughts." As a result, the entire complex was demolished in 1896.

But Chicago's and the federal government's loss was definitely Milwaukee's gain. The buff stone that adorns St. Josaphat's main elevations was unquestionably taken from that site. This selection, quite popular with architects in post–Great Fire Chicago, is the Buena Vista Siltstone. As usual, various architectural historians, including the authors of the 1975 Milwaukee Houses of Worship Survey and the National Register of Historic Places application for this site, have magically transformed this clastic stone into something else quite different, chemically precipitated limestone. While this remains an overt and elementary mistake, the separate practice of calling the Buena Vista a sandstone instead of a siltstone is much more defensible, since its grain size can straddle the boundary between the two classifications.

The Buena Vista Siltstone you see here was quarried in the aptly named but now no longer extant town of Freestone, in southern Ohio. When purchased for the Custom House, it cost, according to an 1878 article in the *American Architect*, from $1.30 to $1.47 1/2 per cubic foot. Very similar in appearance to the Berea Sandstone produced in the northern part of the same state, the Buena Vista is here presented in a variety of finishes, from rusticated to dressed-faced, and from comb-chiseled to reticulated, perforated, and carved patterns that reflect both the whimsy and the skill of nineteenth-century stonemasons. Derived from the Buena Vista Member of the Cuyahoga Formation, the siltstone is Mississippian in age. And, like the Berea and Indiana's Salem Limestone, it possesses the sought-after attribute of being a *freestone*—a rock type that can be reliably cut or sawed in any direction without unwanted breakage.

Though St. Josaphat's massive entrance columns have often been described as taken from the Custom House demolition, too, there's excellent evidence that contradicts this common belief. For a long time I was troubled by the fact that contemporary descriptions and surviving photos of that building do not mention or show columns that resembled the ones here. And then I came across an article in an 1898 issue of the trade journal *Granite* that cleared up the matter and confirmed my suspicion. This report unequivocally states that, unlike the Buena Vista, the shafts were actually obtained from a second Windy City source, the building that then served as Chicago City Hall. While that structure, almost as unsuccessful as the Custom House five blocks to its south, was not completely demolished until 1908, the portico and stairway on its west side were removed a decade earlier to make way for a more accessible ground-level entrance. This was just in time for its columns to be available for the St. Josaphat project. The anonymous *Granite* reporter laid out with both consummate clarity and implied irony the costs involved. The City of Chicago had originally purchased these columns for $2,500 each, for a total of $15,000. Grutza paid only $500 apiece, or $3,000 all told—an 80 percent deduction that might lead the faithful to conclude the good father had divine help in closing the deal. Unfortunately, though, this highly detailed account did not also include an exact identification of the columns' granite, which was simply described as being from Maine. Alas, that great stone-producing state has had several igneous-rock selections from disparate places that resemble the handsome polished selection seen here. Without better provenance I refrain from speculating which of these it really is.

The other geologically derived material visible on the basilica's exterior is the copper sheathing of its dome. When installed in 1992 to replace the original copper, it was described as brown. This suggests that the surface of the metal was then in its oxidized form. However, on recent visits, I've seen patches of green developing, which hint that the dome will eventually harmonize with the rest of the roofing as it returns to its previous copper-carbonate, copper-sulfate patina.

7.5 Forest Home Cemetery

2405 W. Forest Home Avenue
Established in 1850
Geologic features: Lake Border Moraine swell-and-swale terrain; various rock and metal types described below

> **There's a city vast yet voiceless, growing ever, street on street,**
> **Whither friends with friends e'er meeting, ever meeting, never greet;**
> **And where rivals, fierce and vengeful, calm and silent, mutely meet,**
> **Never greeting; ever meet.**
>
> . . . . .

Thus two cities grow forever, parted by a narrow tide,
This the shadow, that the substance, growing by each other's side,
Gliding one into the other, and forevermore shall glide,
Growing ever side by side.

—from "The Silent City," by J. D. Sherwood, as quoted in a
nineteenth-century guide to Forest Home Cemetery

Situated in the swell-and-swale terrain of the Lake Border Morainic System just
north of the Kinnickinnic River, Forest Home Cemetery began as a graveyard
primarily for the congregation of Yankee Hill's St. Paul's Episcopal Church. It
went on over the decades to become Milwaukee's most famous multidenomi-
national burial ground. Initially surveyed by Increase Lapham, the city's great
early naturalist, it now serves as his resting place, as described below. The rolling
topography here is mostly blanketed in *till*, glacial sediment deposited by the
Pleistocene epoch's final, Wisconsin ice sheet. It was Lapham and his successors
down to the present day who've done so much to unlock the complex history of
glacial movement and landscape evolution of the Milwaukee area during and
after our planet's most recent ice age.

Forest Home, besides being the final destination for many of the Cream City's
notable citizens, is also one of Milwaukee's most beautiful open spaces. And it's
an absolute geological treasure trove, too. A full account of the monuments and
geologically derived materials they're made of here would fill an entire book of its
own. I here include eighteen of my favorite sites in the hope this relatively short
survey will inspire you to explore further on your own. Refer to map 7.2 for the
exact locations of the stops cited below.

Several of these sites will acquaint you with Forest Home's plentiful population of faux tree trunks fashioned from America's most commonly encountered architectural rock type, the Salem Limestone. Such fanciful memorials are
usually called *Bedford trees* by modern authors. But they were consistently marketed as *rustic monuments* in the heyday of their production, the 1890s and early
1900s. These custom-carved blocks of southern Indiana biocalcarenite are sheer
Romanticism in rock, heavily laden with nature worship, ruminations on death
and rebirth, utterances of grief and hope, and, to cap it all off, some remarkably
accurate botanical art. The sculptors who crafted them were both sharp-eyed
students of horticulture and skilled purveyors of Late-Victorian schmaltz.

These elaborate grave markers may appear emotionally overwrought or even
creepy to the modern eye, but in their day they were considered tasteful and
heartfelt expressions of familial devotion to the deceased. And their considerable popularity can be measured both by their prevalence here and the fact
they were manufactured in great quantities, in the Hoosier towns of Odon and
Bedford, by such firms as Cross & Rowe, John A. Rowe, Bedford Monumental

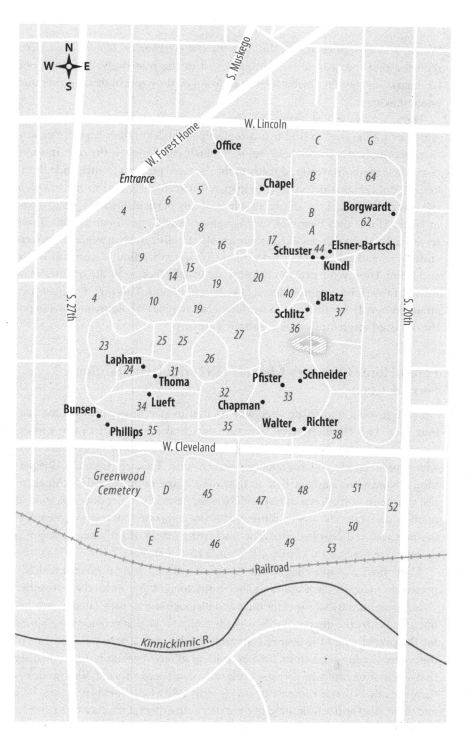

MAP 7.2. Sites in Forest Home Cemetery.

Works, and Correll & Burrell. In 1902 the trade magazine *Monumental News* reported that the last-named company had in the past twelve months done landmark business "necessitating the addition of shop room, steam power and pneumatic tools."

Though by now I've recorded scores of rustic monuments in Midwestern graveyards, I have yet to find two completely identical in their design. I'm sure there are duplicates somewhere, but individuality definitely is the rule. In some cases, their symbolism refers directly to the deceased person's former trade (lumberjack, policeman, ship captain, fireman, railroad engineer, and so on), but their use was certainly not the special province of any particular ethnic group. Here in Forest Home they bear, not surprisingly, mostly German names. But if one travels to Kenosha's Green Ridge, there are also English and Norwegian monikers chiseled in; and at Mount Carmel west of Chicago, you'll find Marubio, O'Reilly, and Leoni. However, rustic monuments are usually lacking in the cemeteries of such ultra-wealthy communities as Lake Forest, Illinois. This might suggest they appealed mostly to middle- and working-class families whose memorializing of the departed was tempered by economic necessity.

Forest Home Cemetery Chapel (1892 with later additions, Ferry & Clas)

On arrival you'll immediately see why it's best to start your Forest Home geo-tour here. This Gothic Revival structure, which also incorporates two conservatories and a crematory, is clad in random-coursed ashlar of Wisconsin's most widely used variety of Lake Superior Brownstone. This selection, the Chequamegon Sandstone, was quarried on Basswood Island in the Apostle archipelago, just northeast of the town of Bayfield. Nowhere else in the city does this somber maroon rock type show off a more interesting assortment of textures, weathering patterns, and crossbedding. The fascinating origin of the Chequamegon is described in the St. Paul's Episcopal Church entry (site 8.9).

Mineralogically speaking, about 75 percent of this stone is quartz, which is present in many of the grains and also in the cement that holds them together. Other constituents include orthoclase and plagioclase feldspars, fated to weather out long before the quartz, and iron-oxide compounds that provide the staining for the particles and the overall coloration of the rock. There are a few bits of mica, too. If you spend some time examining different ashlar blocks, you'll see that while some are *well sorted*—uniformly composed of sand-sized grains— there are others that are somewhat *poorly sorted*, with larger pebbles present here and there. Also notice how the gritty, red decorative mortar makes a nice complement to the stone. There is a fair amount of soot staining on the exterior, but you will still see quite a few examples of unblemished stone faces, or stone newly exposed because of spalling.

Forest Home Cemetery Office Building (1909, Alexander C. Eschweiler)

Completed seventeen years after the Chapel, the Office Building is also faced in random-coursed Lake Superior Sandstone. But is it specifically Chequamegon Sandstone? It's certainly plausible, but there are some other varieties that closely resemble it. Unfortunately, available records don't identify the locale of origin or a quarry name.

Schuster Family Monument (Section 44)

This site is one of the cemetery's largest and best-preserved examples of the "Bedford tree" or rustic-monument type discussed above. Here the characteristic form is expressed: a tree trunk, broken above but still erect, with smaller separate sections in the rear that serve as individual headstones. In this style botanical exactitude is usually observed. On this specimen, however, the sculptor has added a broken, drooping branch that has sprouted weird rootlike extensions that clutch like eagle talons at the bark, furrowed like an American elm's, and at the oak-and-laurel wreath that bears the Schuster name. This nightmarish allusion is balanced on the side by a climbing, cordate-leaved vine, symbolic, one assumes, of eternal life.

But the real treat is found on one of the back headstones, where father Max Schuster, a railroad engineer, is commemorated by a locomotive, complete with cowcatcher, that sprouts out of the base of its trunk section. It's weathered and lichen-encrusted, but still seems to be dutifully chugging along in this sylvan setting.

All that noted, there's some irony that this allegory-ridden graveyard marker is actually formed of a much more ancient graveyard—one that still contains the fossil remains of marine animals that lived in a very different version of the Midwest, 340 Ma before there were any human beings to bury and remember. Look closely at the less weathered portions of the stone and you'll discover a host of shell fragments exposed at the surface.

Kundl Monument (Section 44)

This less ornate rustic monument is included because it presents, at least at time of writing, a cautionary tale: you can't judge a stone's long-term survivability by its popularity or ornamental attributes.

On this Salem Limestone tree trunk, ornament is confined to a cluster of small polypores (bracket fungi), perhaps connoting decay, and two white-oak leaves sprouting from a stumped limb suggestive of the Christian concept of the resurrection. But sadly, the large tablet bearing the names of the two Kundl family members commemorated here has shattered into pieces. A smaller trunk section also lies on the ground, though it's not clear whether it was originally attached to the main stem or not. At any rate, whether this breakage is the result of vandalism or weathering, it illustrates that the porous and relatively soft Salem Limestone is not the best choice if the endurance of a grave-site marker is the chief concern. When mounted on sturdy mausoleum walls the Salem fares better, but in more exposed sites and fragile sculpted forms it's bound to endure only a brief time compared to the granitoid rock types also used extensively throughout the Forest Home grounds.

Elsner-Bartsch Monument (Section 44)

The third of Section 44's Bedford-tree troika is decked out with a floral wreath and a basal herbaceous plant that appears to be some sort of sword fern. But even more indicative of the resurgence and adaptability of life is the thriving community of real mosses and algae that now cover some of the Salem Limestone surfaces. These calciphilic organisms thrive on the calcite (calcium carbonate) that is the natural cement of this fossil-rich sedimentary rock.

Borgwardt Monument (Section 62)

An upright marker simple in design, it's nonetheless one of Forest Home's best examples of Minnesota's Paleoarchean Morton Gneiss. At an age of about 3.52 Ga, the Morton already has an outstanding claim on eternity, and of all the rock types cited in this book it is the stone most monumental, most staggering in its implications, most symbolic of endless time and the ceaseless flow of elements. Here the stone speaks of that flow with alternating, gently undulating bands of dark mafic and light-toned felsic minerals. Taken from the ancient Minnesota River Valley Terrane, this migmatite was a favorite cladding choice of Milwaukee's Art Deco architects (see, for example, the site 5.4 description). But it has also long served as the most beautiful and appropriate of cemetery-stone selections.

Blatz Mausoleum (Section 37; erected by 1897)

A sort of miniature Milwaukee County Courthouse in its insistence on dominating the landscape around it, this most grandiose of Forest Home mausolea has

FIGURE 7.3.   Forest Home Cemetery's Blatz Mausoleum may take the concept of flamboyance to a whole new level, but it also demonstrates how remarkably resistant to weathering Vermont's Devonian-period Barre Granodiorite really is.

been described as Romanesque Revival. I presume this is based on its semicircular entrance arch and rock-faced plinth. But to me it deserves its own special style name, Gilded-Age-Beer-Baron-Ultra-Flamboyant-Super-Deluxe. Everything else has been piled onto it; why not a surplus of adjectives?

Here under the litter of urns and wreaths and acroteria the urban naturalist finds solace in the stone itself. The Blatz clan chose wisely: this is the Barre Granodiorite, Vermont's special gift to the world of monumental stone. I'm not sure how anything so grayly consistent in aspect can be so lovely, but there is some subliminal reason Barre's sedateness works exceptionally well in somnolent settings. Devonian in age and the product of the collision of the microcontinent Avalonia with ancestral New England, this rock type is a granodiorite because most of its feldspar content is in the form of plagioclase. But even to most geologists it's indistinguishable in the field from true granite, and in the building trades it's simply called that instead.

Joseph Schlitz Monument (Section 36; 1888)

The Milwaukee beer baron commemorated here was lost at sea in 1875. Because his remains are not present, this monument, topped with an allegorical female figure pointing heavenward, is in fact a *cenotaph*. However, other family members identified with slant markers are indeed interred around it. Erected thirteen years after Schlitz's death, this Gothic Revival design cost $10,000 and was made of the fine-grained "Blue Westerly" variety of Rhode Island's Narragansett Pier Granite.

While it is no longer quarried, the Narragansett Pier was once a strong competitor of the Barre Granodiorite. Like the Barre, it's a salt-and-peppery igneous intrusive selection that is very enduring and ideally suited for memorials and mausolea. Isotopically dated to 273 Ma, this Middle Permian rock formed in a batholith emplaced at the southern end of the geologic structure known as the Narragansett Basin. The intrusion of magma apparently occurred late in the Alleghenian Orogeny, when compression caused by the collision of eastern North America with Africa had given way to crustal shearing, as what is now southeastern New England rotated counterclockwise.

The superb bas-relief on the front of the monument depicts the ship on which Joseph Schlitz lost his life when the vessel ran aground off the Isles of Scilly. This rendering depicts the SS *Schiller*, named for the great German poet and dramatist, as a graceful three-masted steamer, when in reality she was a chunkier

FIGURE 7.4. A bit fanciful in its details, this lovely carving on the Joseph Schlitz monument nonetheless displays both its sculptor's skill and the admirable qualities of the Narragansett Pier Granite.

two-master. I suspect the stone carvers in Westerly, Rhode Island, did not have an actual picture of the vessel to go by. Nevertheless, they created a nautically accurate stand-in that is a marvel of detail, right down to the ratlines, stays, and choppy seas. For over 130 years the granite has held even the most delicate features with no sign of deterioration.

Emil Schneider Monument (Section 33; ca. 1895)

In the symbology of cemeteries, the frailty of human existence is a common trope, but as things currently lie the Schneider Monument takes this theme one step farther. Here not only the person commemorated has fallen; so has part of the monument itself. The 10-foot female figure that originally stood on a lofty pedestal should now be rebranded as *The Triumph of Gravity*. It reclines headless and on its back in the grass nearby. When I talked to Forest Home staff about this mishap, I gathered it occurred some time ago, and later I did find aerial Google Earth imagery, dated 2011, that shows the detached piece was in the same orientation that long ago.

But at least the geologist can profit from things as they are. The surfaces where the head and one of the hands broke off provide good exposures of unfinished Barre Granodiorite and its crystals of white plagioclase feldspar, glassy gray quartz, black biotite, and traces of silvery muscovite. They seem to have weathered hardly at all in more than a decade. This widely used graveyard stone, discussed in the preceding Blatz Mausoleum description, makes up the entire monument.

Guido Pfister Monument (Section 33; ca. 1891)

In the decorous world of late nineteenth-century graveyard memorials, it is the figure of the somber young woman that is a common vehicle of Victorian sorrow, hope, and consolation. The pensive seated figure here contemplates eternity atop an impressive layer cake of ornament that includes polished pillars with Byzantine capitals. All of it is Narragansett Pier Granite. And like the Schlitz Monument, this work was crafted in Westerly, Rhode Island, for the cost of $10,000. The same sculptor, Joseph Bedford, made the statues for both. Once again the Narragansett Pier demonstrates its ability to retain crisp carved detail over the decades.

Family of Charles S. Richter Monument (Section 33; first half of the twentieth century)

Positioned close to the roadway, the Richter plot consists of an upright marker, adorned with stylized Gothic arches, and a row of slant markers of the same motif and rock type. And a magnificent rock type it is. Once advertised as the hardest and most resilient stone in the monumental trade, it's Wisconsin's own Montello Granite. This was the perfect choice for the Richter family, since Charles had been the longtime owner of the company that quarried it.

The Montello Granite was produced in the Marquette County town of that name from about 1880 to 1976. To understand its special attributes it's necessary to delve into the dramatic details of its origin in the Paleoproterozoic era. From about 1.875 to 1.84 Ga, the Superior Craton, one of North America's earliest building blocks, merged with two other landmasses, a volcanic island arc called the Pembine-Wausau Terrane, and then the Marshfield Terrane, apparently a microcontinent. The result of this geologic equivalent of a three-car pileup was intense folding and the rise of the Penokean Mountains, a lofty range that once spanned the northern part of this and neighboring states. That prolonged episode of collision, crustal crunching, and continent-building was followed by an 80-Ma interval of tectonic calm. Then, suddenly, at 1.76 Ga, all hell broke loose with massive volcanic eruptions in the Marshfield Terrane region. Why this colossal outpouring of searing ash and felsic lava occurred at this particular juncture is not entirely clear, but it might have been triggered by the process of *slab rollback*, where the steepening angle of the subducting oceanic section of the Yavapai Plate caused the adjoining continental crust to stretch and become thinner. This in turn may have caused rifting that encouraged deep-seated magma to rise to the surface. We can find the remains of that great extrusive event today where the igneous rock rhyolite outcrops in the Montello and Berlin regions.

Though it formed from the same mass of rising magma that became the rhyolite, the Montello Granite's portion of the molten mass did not reach the surface before it solidified, as is evidenced by the rock's macroscopic crystal size. Classified as an alkali-feldspar granite, the Montello lacks the plagioclase-feldspar component found in most other granites and is composed, for the most part, of red orthoclase and gray quartz. And there's a lot of the quartz; it can make up 40 percent of the rock. That accounts for its uncommon hardness. There's also a smattering of accessory minerals—magnetite, chlorite, hornblende, and hematite among them. The result is a stone type of striking deep-toned coloration—cherry red to plum purple—and almost supernatural resilience.

When a few years ago I made a personal pilgrimage to Montello to meet with members of its Historic Preservation Society, I was introduced to Bryan Troost, a successor of Charles Richter and the local quarry's last owner. From Bryan I learned the human side of Wisconsin granite production. In this open-hole quarry, I was told, temperatures could peak out at 120 degrees Fahrenheit in the

FIGURE 7.5. The Montello Granite, seen here in Forest Home's Richter Family Monument, was billed as the hardest monumental stone on the market. This ruddy, quartz-rich igneous intrusive rock formed in one of the most violent episodes of Wisconsin's long and event-filled geologic history.

summer and plummet to minus 20 in winter; in either case the work did not stop. And then there was the granite-worker's special occupational hazard, the danger of contracting the serious lung disease silicosis. Some of the rock had to be removed from the headwalls by explosives—at first, black gunpowder and blasting caps inserted in deep drill holes, and then, in the final years of operation, a compound called coalite. But most of the stone was loosened using the centuries-old *plug-and-feather method* of laboriously hand-drilling holes spaced 5 inches apart to the depth of 4 inches. Into these were inserted a pair of shims (the "feathers"), between which a wedge (the "plug") was hammered repeatedly until the pressure built up sufficiently to split even this most resistant of granites. But here too the technology changed over the decades, with the introduction of channeling machines that used mechanical drills or high-temperature flame jets that took the place of human muscle.

Still, after all this effort most of the rock was consigned to the waste pile. Frequently encountered veins of solid hematite and quartz made much of the stone unusable, with only about 10 percent deemed suitable for use. But that which was acceptable had to be further cut to size—and that was done with large-toothed saws set with iron shot. Later, in the 1940s, the saws were replaced by braided steel wire. In most granites, the wire could cut through about 4 feet per hour; but with the Montello, only 8 inches. Surfaces to be polished, such as those you see here, were placed on a cast-iron wheel where they were smoothed with either steel shot or a slurry of water and *emery*, an abrasive containing the very hard mineral corundum and either magnetite or hematite.

All these factors made the Montello one of the priciest granites on the market. Nonetheless, it was widely used for monumental stone, road pavers, riprap, and sometimes, despite its intractable nature, as a beautiful and virtually indestructible building stone. It can still be seen in the last capacity on Chicago's Richardsonian Romanesque Rickcords House, and on the Richters' own Colonial Revival residence in Montello.

Sebastian Walter Monument (Section 33; ca. 1922)

A welcome island of grace and motion in a landscape all too steeped in static dignity, our last example of a Barre Granodiorite monument presents the frequently rendered allegorical woman in a considerably more flowing and even sensuous form. Apparently a southpaw, she stretches energetically to reach a cinerary urn in a niche above. Clearly, we're not in the Victorian period, with its avoidance of the body electric, anymore. But once again the Barre has worn the years well, with its curves, edges, and ornamental details still sharply defined.

Timothy Chapman Monument (Section 33; 1897; Daniel Chester French, sculptor; C. Howard Walker, architect)

In its September 1887 issue the *Monumental News* devoted an article to this memorial, which had been installed here the previous spring. "The sarcophagus," it noted, "is cut from pink Tennessee marble, a marble which is much favored by Boston and other Eastern architects. It stands on a base of pink granite, making a harmonious combination of material."

That harmonious blend of pinkness is not so evident today. The upper portion of the monument is a chalky white; only in a few places does the original rosy tint poke through. The "Tennessee Marble," more accurately known as the Holston Limestone, appears to have altered very noticeably by producing a coating of calcite or other calcium compound that has reprecipitated on the stone's surface. It makes the sarcophagus look as though it's fabricated of true marble. Usually, the Holston in its various patterns and color schemes is reserved for indoor use, and the only other place in the Midwest I've seen it exposed to the elements is the Illinois Centennial Monument on Chicago's Logan Square. There the same variety was chosen, but it has retained its original pink shade without developing the white crust, weathered granular texture, and the community of calciphilic lichens found here. The origin of the Ordovician-age Holston is discussed in the City Hall description (site 5.28).

While its exact identity went unrecorded, the basal rock type appears to be Massachusetts's Milford Granite, more fully described in the site 5.19 account. It's one of the most attractive igneous rock types seen in cemeteries. But it too has weathered considerably, and its normal salmon tone is obscured by soot and other surface contaminants. In contrast the green copper-carbonate/sulfate patina of the presiding angel is a much more attractive form of weathering. This bronze figure is the work of the artist who also sculpted the great Lincoln Memorial statue in Washington, DC.

John Phillips Monument (Section 35)

The most ornate of the rustic monuments described in this book, this robust agglomeration of symbolic curios demonstrates that at Forest Home this headstone style was not exclusively a German American phenomenon. And here the Salem Limestone assumes many forms. On the frontal plaque hanging on the tall trunk is a fine bas-relief bust of the mustachioed Mr. Phillips; below him, a potted calla lily, a species associated with both death and resurrection. Also competing for attention are more potted plants, polypore fungi, a dove, oak branches replete with lobed leaves and acorns, and various flower types, including a morning-glory blossom and what appears to be some species of wild rose. My guess is that the Hoosier stone carver who produced this riot of biologic detail had on an extra pot of coffee the day he did it.

Bunsen Monument (Section 35)

It may be the humblest of the rustic-monument lot, but it's my favorite—a wee weathered stump of Salem Limestone, probably purchased by an immigrant family with little cash to spare, that nevertheless packs quite a botanical punch. Along with climbing English ivy, a token of life resurgent, there are large, cordate leaves that bear the inscriptions (sadly, one is partly broken). Add to that sword-fern fronds, and, remarkably, what appears to be a ribbed barrel cactus tucked into one notch at the base. If this little visitor from desert climes suggests anything more profound than the carver's own collection of houseplants, I don't know what it is. But as an utter non sequitur in a sea of stock images, it deserves a hearty nod of approval.

FIGURE 7.6.   The John Phillips Monument is among the finest of Forest Home's substantial collection of "Bedford trees" made from southern Indiana's Salem Limestone. These symbol-laden memorials were often customized to reflect the deceased person's profession or, as here, to bear his portrait.

Lueft Monument (Section 34)

Whether this "Bedford tree" is another example of the relative fragility of Salem Limestone as a cemetery selection or not is unclear. Historical sources note that some rustic monuments included intentionally broken or detached sections that symbolized a life cut short. Regardless of whether the large grounded piece has always been lying here or not, the fact remains that the standing portion has a well-defined inscription and a well-rendered book of life, as well as the familiar polypore fungi, a potted calla lily, and a floral garland.

Ferdinand Thoma Monument (Section 31; ca. 1894)

Our final rustic monument is a simple affair. Besides the opened inscription scroll the tree trunk is unadorned except for a basal sword fern and, next to it, what appears to be a leaf of sharp-lobed hepatica, a spring wildflower native to the Midwest.

If we look back and consider all the carved plants we've found on Forest Home's Salem Limestone memorials, we see that the sculptors who made them used quite a mixture of models, ranging from indigenous species to cultivated and sometimes exotic horticultural selections. The Mississippian-subperiod rock that is their medium formed some 200 Ma before flowering plants first arose. Only the ancestors of the sword fern were extant earlier; the earliest fern fossils are about 20 Ma older than the Salem itself.

Increase Lapham Monument (Section 24; ca. 1875)

Here we justly honor a remarkable man—a naturalist, geologist, popular-science writer, forestry expert, meteorologist, anthropologist, civil engineer, geographer, and cartographer who on top of everything else was the initial surveyor and designer of the Forest Home grounds. By rights he should have a monument prominently sited and at least the equal in size and durability to that of any merchant or politician here. Instead he rests unobtrusively, along with his wife Ann and other family members, in a shady spot. His own upright marker is a modest affair, made of true marble, albeit of unknown origin.

White marble from Carrara, Italy, and domestic sources as well was an extremely popular graveyard selection here and elsewhere in the middle to late nineteenth century. Unfortunately, its special vulnerability to acidic rainwater and other weathering agents was not well understood. Nowadays many older cemetery markers made of this nonfoliated, calcite-rich metamorphic rock have deteriorated to the point their legends can no longer be deciphered. Increase's headstone has not decayed to that point, thank goodness, though it does show some of the "sugaring" (decomposed, granular texture) that indicates weathering is in fact proceeding apace. It also has quite a colony of calciphilic algae in its lower reaches.

7.6   St. Sava Serbian Orthodox Cathedral

3201 S. 51st Street

Completed in 1958

Architectural firms: Lefebvre & Wiggins; Camburas & Theodore

Geologic features: Fond du Lac Dolostone, Salem Limestone, Copper

Built long after the extinction of Milwaukee County's own building-stone indus-try, this handsome Byzantine Revival church nevertheless is a showplace of our Regional Silurian Dolostone. The ashlar here was quarried not in Lannon, the largest twentieth-century producer, but about 5 miles south of downtown Fond du Lac. Rock-faced and set in random courses, it's best appreciated in this won-derfully open site on a cloudless day, when it glows with hints of tan and yellow. Making up the uppermost crust from south of Chicago to Door County, and from the lower fringe of the Upper Peninsula over to western New York, the Regional Silurian Dolostone began as carbonate sediments on the edge of the Michigan Basin, when this part of North America lay in the subtropics below the equator.

The other rock type on the exterior is that southern Indiana stalwart, the Salem ("Bedford") Limestone. It serves as the dressed-face ashlar and

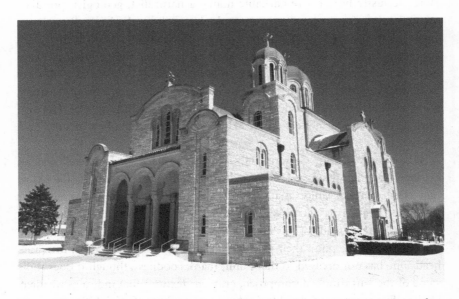

FIGURE 7.7.   The Wauwatosa variety has by no means been the only Regional Silurian Dolostone type used as a building material in Milwaukee County, as the magnificent Fond du Lac Dolostone exterior of the St. Sava Cathedral attests.

columns of the entrance, as well as trim elsewhere. Its grayish-buff uniformity subtly underlines the more complicated textures and tints of the Fond Du Lac Dolostone. And surmounting all, in the grand Byzantine manner, are five domes. Each is sheathed in copper, now nicely weathered to a green, carbonate/sulfate patina.

MILWAUKEE: YANKEE HILL, LOWER EAST SIDE, NORTHPOINT, DOWNER WOODS, AND ESTABROOK PARK; VILLAGE OF SHOREWOOD

8.1 Immanuel Presbyterian Church

1100 N. Astor Street
Completed in 1873
Architect: Edward Townsend Mix
Geologic features: Wauwatosa Dolostone, Potsdam Sandstone, Berea
  Sandstone, Aberdeenshire Granite

In designing this striking variant of the Classic Milwaukee Formula, E. T. Mix displayed his flair for the Italian Gothic. This is most immediately evident in the pointed arches decorated with alternating buff and red voussoirs, and in the vivid underlining provided by the deep-red trim. The dramatic contrast to the lighter tints of the random-coursed Wauwatosa Dolostone is provided by a rock type as rare in this city as the Wauwatosa is common: upstate New York's Potsdam Sandstone. Rare it may now be on the Lake Michigan shore, but in the nineteenth century it had a glowing reputation in the building trades. The August 1891 issue if the *Inland Architect* commented, "Possibly, taking all things together, what is known as the 'Potsdam red sandstone' offers better results than any of the other building stones. At least it has no superior." The journal then went on to note,

> It is a wonderfully durable stone, being composed of pure silica. As to its resistance of fire it is only necessary to state it is used as a lining to furnaces instead of fire brick. It breaks true under hammer and can be sawed, rubbed and carved with the same facility as the best granites. It does not stain, but can be washed off as readily as glass. . . . Frost does

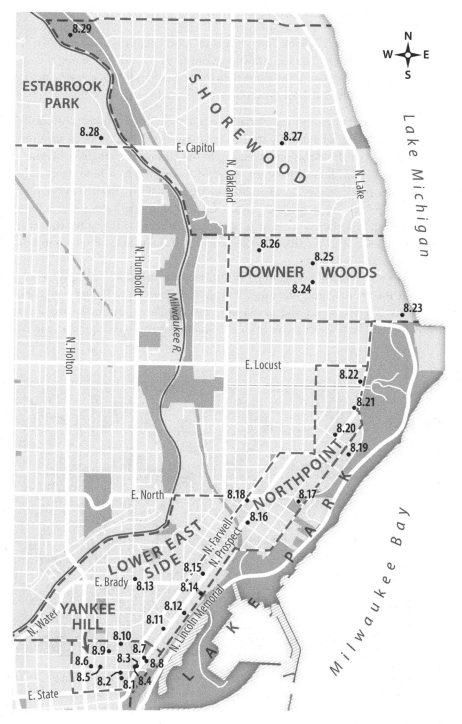

MAP 8.1. Sites in Milwaukee's Yankee Hill, Lower East Side, Northpoint, Downer Woods, and Estabrook Park neighborhoods; and in Shorewood.

not affect it. . . . Somebody once said "God doubtless could have made a better berry than the strawberry, but he never did," so he might have made a better building stone than the Potsdam sandstone, but he never did, or if he did, it hasn't yet been discovered.

Regrettably, the Immanuel Presbyterian's complement of this ruddy clastic sedimentary rock cannot be directly studied at ground level. But it can be admired a little farther up the elevation, especially in the entrance's dark-colored voussoirs. Quarried on the northeastern fringe of the Adirondack Mountains only fifteen miles south of the Canadian border, the Potsdam is Upper Cambrian to Lower Ordovician in age. It was deposited as layers of marine sand not far off the coast of Laurentia, during an episode of high sea level called the Sauk Sequence. Its hardness is due to the silica that cements its hematite-stained quartz grains together.

There's also another, buff-colored sandstone present, but it's been given short shrift in the architectural literature. It is most visible in the lighter voussoirs, archivolts, buttress flaps, and associated trim. While it's distinctly bad form to remove any still-intact stone from a building, I have found quite a crop of

FIGURE 8.1. Detail of the upper façade of the Italian Gothic Immanuel Presbyterian Church. Here the main exterior stone, rock-faced Wauwatosa Dolostone ashlar, is complemented with both red Potsdam Sandstone from New York State and what is probably Ohio's Berea Sandstone.

spalled-off pieces lying on the ground. These have had a tint and gritty texture that almost certainly identify them as northern Ohio's Berea Sandstone, a very popular choice of Milwaukee architects. The other possibility, the look-alike Buena Vista Siltstone, lacks the slightly abrasive feel. For more on the Berea see the site 5.15 account.

The final stone type of note on this geologically diverse house of worship is much more accessible to the hand lens–wielding urban geologist. The gleaming shafts of the entrance columns are fabricated of what architects of Mix's time called "Scotch Granite." This is almost always a synonym for Aberdeenshire Granite, and given the columns' color there's a very good chance they're examples of the "Red Peterhead" variety quarried in the northeastern Scottish city of that name. It's a common sight at the entrances of nineteenth-century American churches and courthouses, where, as here, it was almost always employed as polished pillars. Like the Berea Sandstone, the Aberdeenshire Granite is Devonian in age. Make sure you take a lingering look at its handsome, medium-grained mixture of red orthoclase feldspar, black hornblende, and gray quartz.

8.2   James S. Brown Double House

1122–1124 N. Astor Street
Completed ca. 1852
Architect: Unknown
Geologic feature: Cream City Brick

Its Federal style signals that the Brown House is a survivor from Milwaukee's early decades, when this high ground of Yankee Hill, now so prestigious, was a distant plot of real estate far from main settlement areas. And indeed this paired residence is considered one of Milwaukee's oldest surviving buildings. While it was subjected to substantial heretical alterations over the years, it was beautifully restored after the turn of the millennium.

From the standpoint of geologically derived building materials, the Brown House offers us the earliest nonecclesiastical use of locally manufactured Cream City Brick cited in this book.

8.3   Jason Downer House

1201 N. Prospect Avenue
Current name: Milwaukee Art Museum Research Center

Completed in 1874
Architect: Edward Townsend Mix
Geologic feature: Wauwatosa Dolostone; Cream City Brick

The finest Gothic Revival residence in the city, the Downer House is the perfect illustration of the Third (and best) Version of the Classic Milwaukee Formula, where architects have used a harmonious blend of our two locally derived building materials to create a light-toned play of contrasting colors and textures. Here the Wauwatosa Dolostone, in both rock- and dressed-faced finishes, occupies its customary role as the building's plinth, perforated here and there by basement windows. This site, including the stone you see here, was originally intended for an Episcopal church, but the Wauwatosa turned out to be just as apt in this more domestic application. Above its base of palest gray rises the light yellow of the Cream City Brick. Nowhere else is this simple pairing more effective.

FIGURE 8.2. Two of Yankee Hill's most geologically splendid buildings stand like proud sisters on adjoining lots. At right, the Jason Downer House, an exemplar of the Classic Milwaukee Formula, is a harmonious blend of Wauwatosa Dolostone and Cream City Brick. In contrast, the George P. Miller House combines a much less common rock type—apparently Minnesota's Hinckley Sandstone—with Cream City Brick and terra-cotta.

8.4   George P. Miller House

1060 E. Juneau Avenue
Current name: Junior League of Milwaukee
Completed in 1887
Architect: Attributed to George Fiedler
Geologic features: Hinckley Sandstone, Cream City Brick, Terra-Cotta,
   Copper, Wrought Iron

When first I spotted this amazing house from some distance away, I told myself that its lower, rock-faced portion had to be another Milwaukee application of the "Kasota Stone" variety of Oneota Dolostone. At that point, I'd not seen any other architectural rock type with that distinctive pinkish-yellow color. Little did I realize that the Miller House rock, far from being a common and easily identified type, would prove to be one of the city's most mysterious.

Checking the historical sources, I found that the selection here was usually described as "Abelman stone" or "Abelman quartzite." These names trigger an immediate association for any person acquainted with Wisconsin geology: they seem to refer to Ableman's Gorge, in Rock Springs and on the northern flank of the Baraboo Range. This holiest-of-holies of Upper Midwestern geology is a renowned site visited over the past century by hordes of Earth science professors and their students. There they dutifully study the Van Hise Rock, a remnant of steeply dipping bedrock that beautifully illustrates important principles of structural geology and metamorphic petrology. Also of interest in the gorge are quarries past and present that expose both the Paleoproterozoic Baraboo Quartzite, some of it with vertically tilted ripple marks that march up a headwall, and the overlying Upper Cambrian conglomerate and sandstone.

But, as was pointed out to me by Esther Stewart, a sedimentary rock expert at the Wisconsin Geological and Natural History Survey, historians always spell the name of the Miller House stone slightly differently than the gorge's—Abelman versus Ableman. While this suggests to me that the former refers to some other location, I nonetheless did my best to hunt down Ableman's Gorge rock that could fit the bill here. I knew it couldn't be that locale's Baraboo Quartzite; it's normally deep red, plum purple or purplish gray. And, as Wisconsin building-stone expert Ernest Robertson Buckley noted in 1898, "it is one of the most refractory of stones and dresses with exceeding difficulty." This horrendously hard rock has been used as railroad ballast and street pavers, but it's not a likely medium for the fancy carving on the Miller House's façade. However, there is also Cambrian sandstone that overlies it in the gorge's quarries. Unfortunately, that isn't a particularly good match, either. To once again quote Buckley, it too can be rather "refractory" and is "often difficult to dress." It was usually employed instead for coarser applications: bridgework, riprap, and foundations.

However, as is often the case, I later came across yet another source that broke my mental logjam. Russell Zimmermann's *Magnificent Milwaukee* (see Selected Bibliography) describes the Miller House rock as "Minnesota sandstone" instead. That made instant sense. There is a selection quarried in that state, the Hinckley Sandstone, that's a dead ringer for the Miller exterior. As a matter of fact, back in 1880 Minnesota geologist Horace Winchell noted that this rock type closely resembles the "Kasota Stone" in color, which accounts for my own first guess. The Hinckley was used in a number of significant Midwestern buildings of that era, including architectural landmarks on the campuses of the Universities of Minnesota and Illinois. And when I checked with Precambrian specialist Amy Radakovich Block at the Minnesota Geological Survey, she and her colleagues Julia Steenberg and Terry Boerboom were quite sure, based on the photos I'd sent them, that this is indeed the correct identification.

The Hinckley Sandstone, quarried about 70 miles southwest of Duluth until 1976, has despite its lighter tint something in common with the Lake Superior Brownstone varieties of northern Wisconsin and Michigan's Upper Peninsula. Like them, it was deposited in the Midcontinent Rift, a giant rent in what is now North America's midsection. This vast expression of dramatic tectonic activity, and the massive outpouring of lava that accompanied it, may have been caused by an unusual combination of crustal stretching and hot-spot activity occurring in the same region at 1.1 Ga. At any rate, once the mayhem had ended, sediments washed down from nearby highlands began to accumulate in this huge trenchlike depression—the sand that would become the Hinckley formation among them. An isotopic study of the small comple-ment of zircon grains contained in this rock suggests that its age is either latest Mesoproterozoic or very early Neoproterozoic. Sometimes described as an orthoquartzite rather than a sandstone, it's composed of quartz grains securely cemented by silica.

While the presence of Hinckley rather than Ableman's Gorge sandstone on the Miller residence is likely but not completely confirmed, the identities of the house's other geologically derived materials are much more certain. On the second-floor exterior, there is Cream City Brick, a little sooty, framed by stone trim and carved ornament. Above that, a belt course of brown terra-cotta; and above that, green-patined copper that serves as another belt course and a dormer surround.

Also of note is the black-painted wrought ironwork of the entrance and frontal terrace. Unlike cast iron, which has a high carbon content and must be thoroughly melted and then set in molds to achieve the preferred shape, wrought iron is less brittle owing to its reduced carbon level. It is heated just to the point of malleability, where it can be hammered or rolled into the desired form.

8.5   All Saints' Episcopal Cathedral

828 E. Juneau Avenue
Original name: Olivet Congregational Church
Completed in 1868
Architect: Edward Townsend Mix
Geologic features: Wauwatosa Dolostone, Cream City Brick

This Gothic Revival church is another exemplar of the Third Version of the Classic Milwaukee Formula featuring both of this county's locally derived building materials. But here the basal Silurian Wauwatosa Dolostone is present only as a very low plinth. The Cream City Brick dominates.

8.6   Summerfield United Methodist Church

728 E. Juneau Avenue
Original name: Summerfield Methodist Episcopal Church
Completed in 1904
Architectural firm: Turnbull & Jones
Geologic features: Buena Vista Siltstone, Salem Limestone

Any major buff-colored stone building constructed in Milwaukee from the beginning of the twentieth century onward is most likely decked out in Salem Limestone. In that regard, such landmarks as the Public Service Building (site 6.2) and the First Church of Christ, Scientist (site 8.11) are good examples of the dominance of that southern Indiana carbonate rock. Still, there are exceptions, and the Summerfield Methodist is one of them.

That's not to say that this church is devoid of the Salem. It's present as trim and archivolts. But the main expanse of buffitude is another Mississippian-age freestone, the Buena Vista Siltstone. It's the same Ohio-quarried clastic rock used, in recycled form, in the St. Josaphat Basilica (site 7.4). Here it can be further distinguished from the Salem by its warmer tint, general lack of fossil fragments, finer texture, and rock-faced finish. For more on the Buena Vista Siltstone see the St. Josaphat section.

8.7 Robert Burns Statue

1249 N. Franklin Place
Installed in 1909
Sculptor: William Grant Stevenson
Architect: Julius E. Heimerl
Geologic features: Nictaux Granodiorite, Bronze

On a scale from 1 to 10 rating the relative rarity of building stones in Milwaukee, the rock forming the plinth of this oxidized-bronze statue of the great Scottish poet might just warrant an 11. If it exists anywhere else in this city, or anywhere else in the American Midwest, I yearn to know of it.

Historical sources and public-sculpture guides call the stone "pink Nova Scotia granite." In fact, this fine- to medium-textured igneous rock is light gray to buff, depending on the lighting, with only the slightest subliminal hint of pink. (The way architects, sculptors, and stone merchants perceive color in stone, and usually inflate grays into overt blues or pinks, is a tribute to their artistic imaginations, or marketing skills.)

During an extensive review of granitoid rocks quarried in that Maritime Province, I found some remarkably detailed accounts written by Canadian geologists. Happily, these led in the end to a single, very plausible candidate for the stone here: the Nictaux Granodiorite. The other choices were too coarse, the wrong tint, or unavailable at the time of this monument's creation. The Nictaux, it turned out, was highly regarded for monumental uses such as this, and was actively quarried from about 1889 to 2007. Photographic plates of it in volume 2 of William A. Parks's 1914 *Report on the Building and Ornamental Stones of Canada* match the rock here very well indeed.

Quarried near the eponymous town, the Nictaux Granodiorite is part of the Devonian-period complex of granitoids intruded into Meguma. This wandering terrane originally was part of Gondwana, but ultimately it collided with the Avalonia section of what is now eastern Canada to form the southern portion of Nova Scotia. The Nictaux is composed mainly of off-white plagioclase feldspar, gray quartz, and black biotite mica. Another of its identification traits is the occasional occurrence of *knots*—larger dark clots of the biotite. See if you can spot any here.

The final mystery is why a stone this achingly uncommon was chosen for this site. If it was to make some obscure allusion to Robbie Burns's homeland—after all, Nova Scotia is Latin for New Scotland—why not go the full transatlantic distance and choose the much more frequently used Aberdeenshire Granite instead? It can be seen a few blocks away at the Immanuel Presbyterian Church, and at Juneau Town's Mackie and Mitchell Buildings as well. But whatever the reason, the Nictaux's presence is an unexpected delight to the rockhound and still more evidence of the Cream City's outstanding geologic diversity.

8.8   Exton Apartments

1260 N. Prospect Avenue
Completed in 1939
Architect: Herbert W. Tullgren
Geologic feature: Salem Limestone

The quintessence of 1930s stripped-down sleekness, this bay-windowed beauty shares with the Mariner Building (site 5.20) the distinction of being a perfect expression of Salem Limestone in the Art Moderne mode. In earlier decades this Mississippian-age carbonate rock from southern Indiana was used to communicate either rock-faced ruggedness or intricate neoclassical detail. Here, however, its uniform tone and smooth-sawn texture enhance the intrinsic appeal of uncluttered surfaces.

8.9   St. Paul's Episcopal Church

914 E. Knapp Street
Completed in 1883
Architect: Edward Townsend Mix
Geologic features: Chequamegon Sandstone, Bushkill Slate

On one of my daring forays into the savage wilds of Yankee Hill, when I stopped to admire this darkling expression of the Richardsonian Romanesque, a church staff member spotted and watched me intently as I examined his building's exterior a little too enthusiastically. In Milwaukee as in other urban places, geologists are a rarely encountered species, and the staffer's concern about the sanity of a stranger staring at and photographing small sections of time-weathered stone was completely understandable. Especially so, since I was audibly discussing the fine points of sedimentary petrology with myself as I scrutinized the pebbles embedded here and there in the matrix of smaller sand grains. But eventually curiosity overcame caution, as it often does in this city, and my observer asked me what I was doing. This led to a long conversation and an impromptu guided tour of the church's magnificent interior. Then other officials and docents got involved; and in the next few days I was handed and emailed a boatload of information on St. Paul's history and design. So I can certify that this is a house of worship blessed with a congregation both proud and welcoming.

What had transfixed me so thoroughly on the outside was St. Paul's splendid exposure of Chequamegon Sandstone. Here far-northern Wisconsin's dramatic geologic past has come to Milwaukee. This classic Lake Superior Brownstone variety was quarried a full 300 miles to the north-northwest, on Basswood Island in the pine-studded archipelago of the Apostles. In the 1880s brownstones were all the rage architecturally, and various production sites in Bayfield County sprang up as regional stone companies competed, quite successfully, with New England's famous Portland Sandstone.

Like the other main type of Lake Superior Brownstone found in Milwaukee, the Jacobsville Sandstone of Michigan's Upper Peninsula, the Chequamegon is as temporally elusive as it is duskily beautiful. Lacking fossils and other features that stratigraphers depend on to definitively assign a rock unit to a particular

FIGURE 8.3.   St. Paul's Episcopal Church is Milwaukee's most impressive example of Chequamegon Sandstone, the Lake Superior Brownstone variety quarried in northernmost Wisconsin's Apostle Islands. But it's also home to an unusual roofing slate that was quarried in Pennsylvania.

period, experts have tentatively placed it at various spots on a half-billion-year slice of the geologic time scale. Indeed, it could be as old as the late Mesoproterozoic or as young as the early Cambrian; somewhere in the Neoproterozoic, between those two end points, is probably the most likely. But regardless of its age, we know its deposition followed the cataclysmic Midcontinent Rift event in which ancient North America almost split apart along a curving line of rampant volcanic eruptions. That great breach in the Earth's crust passed through what is now northwestern Wisconsin and Lake Superior, and the thick layers of basaltic lavas it produced were subsequently overlain by the younger Chequamegon strata.

St. Paul's also boasts another rock type of great interest. Less immediately noticeable but also considerably less common in modern Milwaukee, it's easiest to locate on the turret at the church's northwest corner. As you'll see if you look closely, the black stone-shingle roofing there has a distinctive weathering pattern of diagonal white streaks or "ribbons." While I've found no documentation confirming or denying my assignment, this feature is usually a dead giveaway for eastern Pennsylvania's Bushkill Slate, also known in the trade as "Chapman Slate." Ordovician in age and hence younger than the Chequamegon Sandstone beneath it, this fissile and foliated metamorphic rock began as *flysch* (deep-sea mud) deposits in a forearc basin situated between Laurentia and an elongated cluster of volcanic islands. In those days the Atlantic Ocean did not exist, but a precursor, the Iapetus Ocean, still did, though it was closing as the island arc approached the continent.

8.10   First Unitarian Church

1342 N. Astor Street
Completed in 1892
Architectural firm: Ferry & Clas
Geologic feature: Salem Limestone

This Gothic Revival church, just one diagonal block from St. Paul's, is only nine years younger than the latter. But in that short span tastes in building stone, and in building-stone color, had largely changed. As Wisconsin geologist Ernest Robertson Buckley put it not long after the First Unitarian was erected, "Until a few years ago brown stone was all the rage, . . . but the eye became weary of gazing at long rows of somber colored buildings, and the fashion changed to light colored stone, where it now rests, awaiting the next reversal." And the Salem Limestone, produced in immense quantities in Indiana and shipped by rail at reasonable prices, was the primary rock type to effect the transformation from dark to light.

While the Salem is by no means an uncommon sight in Milwaukee—in fact, it's found practically everywhere—this church's ashlar facing Ogden Avenue is one of the city's most superb examples of why this rock is a *biocalcarenite*, a limestone largely composed of fossil fragments. Some of the blocks are as coarsely textured as the Salem gets, and are absolute riots of ancient life frozen in time by their calcite binder. Among the identifiable forms of the Mississippian-subperiod marine fauna are bryozoans, crinoid stems, and brachiopod valves. Less visible but no doubt present are nit-sized *Globoendothyra baileyi* foraminifera.

8.11   First Church of Christ, Scientist

1451 N. Prospect Avenue
More recent name: 1451 Renaissance Place
Completed in 1909
Architect: Solon S. Beman
Geologic feature: Salem Limestone

Until recently serving as a venue for weddings and other events, this erstwhile house of worship was the work of the noted Windy City architect who also designed the Pullman community on Chicago's South Side, and other Christian Scientist churches as well. It's yet another exposure of the Salem Limestone, this time in a restrained, dressed-faced, neoclassical guise. The best place to examine its fossil-fragment texture is on the shafts of its Doric columns.

8.12   Charles L. McIntosh House

1584 N. Prospect Avenue
Current name: Wisconsin Conservatory of Music
Completed in 1904
Architect: Horatio R. Wilson
Geologic features: Jacobsville Sandstone, Galesburg Paving Brick

As discussed in the First Unitarian Church section (site 8.10), the last two decades of the nineteenth century witnessed a profound change in architects' building-stone preferences. Brownstones, so consonant with the Richardsonian Romanesque style, were largely replaced by light-toned rock types considered more appropriate for the neoclassical hauteur of the Beaux Arts movement. Dark,

FIGURE 8.4.   In combining a preferred stone of one era with the style of the next, and in its offbeat ornamental use of paving brick, the Charles L. McIntosh House is a beautiful Beaux Arts oddity bewitching to geologists in search of the unfamiliar.

iron-rich sandstones continued to be used in some relatively modest and distinctly nonclassical residential projects, such as the Crane, Kern, and Wiswell Houses featured later in this chapter. But for anything as grand and becolumned as the McIntosh manse, pale gray was a sure bet to prevail. A liberal dose of Salem Limestone would be proper, seemly, and expected, given the refined lines and aristocratic dialect spoken here.

Instead, we have the edifying shock of a style transgressed. The building seems to be a negative photographic image of itself, incongruous, wonderful. Roman civic dignity is clothed in barbarian reds and maroons, the elegance of sunlit order seen at midnight. Even when the total mixed message is deconstructed into its two primary building materials, the magic persists.

Stand close to the stone of the Composite-order columns of the portico, or in front of the door or window surrounds, and you'll see the rock there is peppered with even darker spots. This is the aptly named "Raindrop" variety of the Jacobsville Sandstone, produced in Marquette, Michigan. The fanciest of the Lake Superior Brownstone clan, it does indeed look as though it has just been lightly wetted by a passing rain shower, or perhaps by a rogue lawn sprinkler. This pleasing oddity of pattern made it much sought-after by purveyors of the curious and sublime, mostly in Richardsonian applications. Like the others of its kind, the Jacobsville was deposited in the Midcontinent Rift at some point in the half-billion-year gulf of time from the latest Mesoproterozoic to the Cambrian.

On the other hand, the main elevations are clad not in rock but in the most unlikely form of fired clay, designed for a much different use: paving brick of the red to pink Galesburg brand. It was manufactured in the Illinois city of that name, which, besides producing the poet Carl Sandburg, was a center of a thriving industry that turned out millions of bricks for America's streets. Galesburg, situated in the northwestern reaches of that great downwarped structure known as the Illinois Basin, sits atop Pennsylvanian-subperiod strata. These include deposits of shale with a composition ideal for fabricating very hard pavers not easily worn down by heavy wheeled traffic.

All this leads back to the main dilemma. Why was dark rock with even darker spots chosen for neoclassical columns and trim? Why do road pavers run up vertical walls? Why is the whole house such a magnificent whatzit? Was it just a matter of grabbing the materials most convenient or affordable at the time of construction? Or were the architect and owner thumbing their noses at all the surrounding Prospect Avenue propriety? Or did they just come from an alternative universe where Beaux Arts in Brown was normal?

8.13   St. Hedwig Catholic Church

1702 N. Humboldt Avenue
Completed in 1887
Architect: Henry Messmer
Geologic features: Wauwatosa Dolostone, Cream City Brick

Now part of the Three Holy Women Parish, St. Hedwig Church is a slight variation on the Third Version of the Classic Milwaukee Formula, which combines a Wauwatosa Dolostone plinth with elevations of Cream City Brick. The variation manifests itself in the inclusion of some red clastic rock type as façade trim. It resembles the "Portage Red" brand of the Upper Peninsula's Jacobsville Sandstone, but I've found no mention of it.

Unlike many of its sister churches of this era, St. Hedwig has a brick exterior that was given a good cleaning in the 1960s, and there seems to have been little new accumulation of grime since. This may very well be architectural evidence of the federal Clean Air Act, initially passed in 1963 and further amended in the years following. Various studies have confirmed that implementation of this legislation has led to a dramatic decrease in air pollution. Nowadays it's difficult for inhabitants of American cities to comprehend just how toxic urban air once was. But many older buildings less cared-for than this one still bear silent, sooty witness to the way things used to be.

8.14 Compass Statue

1724 N. Prospect Avenue
Completed ca. 2005
Sculptor: John Barlow Hudson
Geologic features: Wausau Granite, Athelstane Granite, Mellen Gab-
bro, Stainless Steel

Like the Tekton Series Sculpture (site 6.24), this art installation seems to have been designed specifically for rockhounds and geology-class field trips. Sculptor Hudson is to be roundly praised for celebrating and displaying some of the Badger State's most famous ornamental-stone types. Situated along the path leading down from Prospect Avenue to the pedestrian bridge over Lincoln Memorial Drive, it consists of a quartet of blocks of Wisconsin-quarried igneous rocks that have figured prominently in the monumental and architectural stone trades. They are set in a sort of compass rose connected by curved tubes of stainless steel. As we'll see, that shining metal has its own geologic legacy as well.

All four mounted stone blocks are unpolished and look much as they would have as they left their quarries. And each contains drill holes made as part of the rock-splitting process. The positions of each specimen are as seen from the sculpture's western side, facing lakeward.

The uppermost specimen, at twelve o'clock, is the "Ruby Red" variety of Wausau Granite. This stone, more fully discussed in the 5.41 and 5.43 site descriptions, has been dated to 1.835 Ga. An alkali-feldspar granite, it contains very little plagioclase feldspar, but where it is present, it's visible as white oligoclase. Much more prevalent are the crystals of red microcline, black biotite mica, and gray quartz.

The second, three-o'clock rock, is listed by the sculptor as "Glacial Rose," a variety I can find no other reference to. But it appears to be Wausau also, and may be equivalent to its "Sierra Rose" variety. On its eastern face you'll see diagonal bands of biotite that would almost certainly have precluded this section being used for a fancy monument base or cemetery headstone. There also seems to be a xenolith or dike of lighter-colored rock running through it.

The stone at six o'clock forms the base that holds up the rest. Cited as "Amberg Silver Grey," it's a brand of the Athelstane Granite apparently synonymous with "Wisconsin Silver Grey." I'm not sure why both the artist and quarrier prefer to use the British spelling of its shade, but writing as a red-blooded American author unwilling to bow my head to transatlantic monarchical influences, I can attest it is truly gray. Like the Wausau, the Athelstane is Paleoproterozoic in age. Its particulars have already been delineated in other site descriptions, including that for the Federal Building (5.6). Still, what makes this specimen so notable is that it's my one documented Milwaukee example of the Athelstane's fine-grained form quarried, as advertised, in the Marinette County town of Amberg. More than half of its composition is a blend of orthoclase and microcline feldspars (white and light gray). Darker-gray quartz constitutes most of the rest, though some black biotite mica is present, too.

However, the most significant rock of the lot is at nine o'clock. This is the Mellen Gabbro, often misidentified as a granite when in fact it's an igneous intrusive rock of an entirely different character and classification. It was produced by the cooling of mafic rather than felsic magma. The coarse-textured stone, which contains plagioclase feldspar and ferromagnesian minerals, may here seem a rather nondescript dark gray, but when it's polished, it is a striking, Stygian black with glittering, labradorescent crystals that resemble those of the Larvikite Monzonite featured in the site 6.9 section. So highly regarded was the Mellen in the early twentieth-century heyday of its production that it was used as cladding on such Chicago Art Deco masterpieces as the Carbide & Carbon and Field Buildings. It's almost certainly present in that guise in Milwaukee, too. But documentation as to where, exactly, is lacking.

The Mellen Gabbro, sadly no longer produced for architectural use, was quarried in far-northern Ashland County. It owes its existence to the formation of what geologists dub the "MCR," the Midcontinent Rift. This amazing feature, described more fully in chapter 2, occurred during an episode of immense crustal extension and tectonic activity that almost tore our continent apart. And the Mellen was one product of that rifting.

While the Compass Sculpture's rocky components may steal the show, we mustn't forget that the stainless steel that holds them together is just as geologically significant as they. At the core of its story is the mining of iron, largely extracted in modern times from Banded Iron Formation deposits. (For more on this, see the site 6.9 discussion.) While the weathering and enameled forms of this anthropogenic metal have already been covered at sites 6.1 and 6.9 respectively, the stainless variety deserves its own mention because it remains so anomalously rust-free in this highly corrosive environment of acidic rainwater and an oxygen-rich atmosphere. Its miraculous inertness in the face of such implacable enemies is due to the fact it isn't what steel normally is—a mixture of iron and carbon. Rather, the second of these elements has been replaced by chromium; nickel and molybdenum sometimes play supporting roles as well. But it is primarily the substitution of chromium that results in the gleaming, unsullied super-alloy you see here. Not surprisingly, though, attractiveness and durability come at a price, for the sole ore of chromium, predictably named chromite, is by no means as common as the widely distributed iron ores hematite and magnetite. In fact, the world's supply of commercially viable chromite is mostly restricted to areas like South Africa's Bushveld Complex, where Precambrian ultramafic rocks, formed from mantle-derived magma, are now exposed at the surface. So the stainless steel you see here can be justly regarded as a gift fashioned from the Earth's distant past and deep interior.

8.15   Sarah and Charles Allis House

1801 N. Prospect Avenue
Current name: Charles Allis Art Museum
Completed in 1911
Architect: Alexander C. Eschweiler
Foundation: Piles
Geologic features: Ohio Brick, Lake Superior Brownstone, Bronze, Numidian Red Breccia, Pinoso Limestone, Carrara Marble, Hauteville Limestone, Chiampo Limestone, Rosso Ammonitico Veronese Limestone, Imperial Porphyry, Inlaid Stone

While it has only a small fraction of the floor space of some other Milwaukee museums, this Tudor Revival mansion has a remarkable diversity of geologic detail. It's fairly bursting with building and ornamental materials. So rich is the Allis's inventory of documented brick and stone types that it's best to present them by their exact locations on and in the house. The following identifications are based on information obtained from the Allis Museum's helpful staff, their excellent and detailed website, and the description in Richard Perrin's *Milwaukee Landmarks* (see Selected Bibliography).

Foundation and Exterior

The first item of interest is this site's foundation, a particularly solid one for a relatively small building that was designed as a private residence. Reportedly the Allis House is set on piles driven to a depth of 50 feet. Records of deep wells drilled just half a block away reveal that in this locale the bedrock begins much farther down, at 140 feet. So the Allis piles terminate in a thick overlying layer of clay instead. This clay is not stiff, dewatered hardpan, which here isn't encountered until 125 feet. Nevertheless it provides sufficient anchorage—and not only here, but also for some of Milwaukee's much larger and heavier structures set on piles.

The house's aboveground portion is no less impressively constructed. Handsome, rough-textured facing brick is the material of choice for most of the building's outer surfaces. Described by Perrin more than half a century ago as mauve, it could still be said to have that tint in places. But now it's generally something closer to shades of standard medium brown, especially when seen from across the street. All records indicate these bricks were manufactured in Ohio, but no surviving reference details exactly where therein. Historically, the Buckeye State has been one of the country's major producers of fired-clay products, with its northern firms generally using clay from Pleistocene glacial deposits, and those in the unglaciated southeast taking advantage instead of their local Pennsylvanian-subperiod bedrock with its shales and underclays.

The Allis exterior also has a generous helping of Lake Superior Brownstone at the base of the center section, and in the balcony, entrance, and window surrounds. Note that it has been intricately carved in places. Once again, available documents do not specify the locale of origin, but almost certainly the rock is either Chequamegon Sandstone from the Apostle Islands region or Jacobsville Sandstone from Marquette or L'Anse, Michigan.

First-Floor Interior

Enclosed Porch. While visitors now access the house through its western wing, this space, with its bronze door, was the original entrance. Its walls are clad in more Lake Superior Brownstone.

Marble Hall. According to the museum's website, this central space contains four Italian marble types installed by Italian craftsmen. But if Perrin's list is correct, and it appears to be, only two of these stones were actually quarried in Italy, and only one is really a marble. That isn't to say the museum authorities are necessarily wrong in a broader sense: after they were obtained from their disparate countries of origin, these rock types may have all been prepared in Italy and then shipped together from there to Milwaukee. Not only does *la bella Italia* have a long history of supplying *scalpellini* (stonemasons); she has long served as the world's stone-finishing center. Starting with the flooring, the pavers appear to be the "Florido Crème" variety of Spain's Pinoso Limestone. Also well known as "Crema Marfil," this off-white, black-veined carbonate rock dates to the Eocene epoch and is still produced just west of the seaport city of Alicante. Next to it you'll find the white and heavily silver-veined stone of the baseboards. This isn't cited in any source but is almost certainly a variety of Italy's Carrara Marble—probably either "Pavonazzo" or "Calacatta." The illustrious Carrara, quarried since ancient Roman times in Tuscany's Apuan Alps, is a genuinely metamorphic type that began as a Late Triassic to Early Jurassic limestone protolith and was subsequently transformed by tectonic activity into true marble in the Oligocene and Miocene epochs.

The Marble Hall's main wall panels are splendid examples of one of the classiest of polishable carbonate rocks, the yellow-tan "Hauteville Fleuri" brand of the Hauteville Limestone. Cretaceous in age, it comes from eastern France, not far from the Swiss border. Though it's definitely understated in comparison to the other, more highly patterned selections in this hall, it can sometimes contain remains of Tethys Ocean marine life, including ammonite shells and signs of bioturbation caused by burrowing organisms. It is here spiced up by showier pilasters and trim of what apparently is the "Tavernelle Clair" variant of Italy's Eocene Chiampo Limestone. If this assignment is correct, this is a particularly brecciated version of the Chiampo, with a good deal of silver-gray infill between the buff clasts.

Marble Staircase. The treads and risers are Pinoso Limestone continued from the Marble Hall's flooring, once again flanked with a veined Carrara Marble baseboard and Hauteville Limestone wall paneling.

Dining Room. A geologic high point of this stone-rich site can be found in the hearth pavers and trim of this room's fireplace. This is Numidian Red Breccia, known in the building trades as "Brèche Sanguine." And sanguine this clastic sedimentary rock truly is. Composed of a jumble of angular limestone chunks held in a matrix richly suffused with the mineral hematite, this selection is called a "marble" by architects because it is relatively soft and takes a high-gloss finish. One of the most striking of the Mediterranean Basin's vast array of beautiful building stones, Numidian Red Breccia is Jurassic to Cretaceous in age. It was quarried from the later 1800s to 1950 in the northern Algerian coastal highlands, between Kristel and Sidi ben Yebka.

Library. The fireplace mantel is made of the museum's lithic leitmotif, Lake Superior Brownstone. Also of interest in this room is the sculpture of lions— a common Allis House theme—sitting on a pedestal of more Numidian Red Breccia. Purchased from Eugene Glaenzer & Company of Venice, this leonine composition was fabricated of yet another famous sedimentary rock often marketed as a marble—the Rosso Ammonitico Veronese Limestone. Understandably, Italian geologists often just abbreviate it to "RAV."

Jurassic in age, the RAV has been quarried since Imperial Roman times near the northern Italian metropolis of Verona. It generally comes in two colorations, one of which looks rather like pink gnocchi floating in tomato soup, and another

FIGURE 8.5. Each room of the Allis House has its own inventory of geologic treasures. Here, in the library, a sculpted group of lions, carved from northern Italy's Rosso Ammonitico Veronese Limestone, sits on a stand of Numidian Red Breccia.

with a golden-brown, mottled-butterscotch look. The latter, known as "Nembro Giallo di San Ambrogio," seems to be the selection used here for this carving. The RAV, which was originally deposited in the great Tethys Ocean that once stretched from western Europe to southeast Asia, contains abundant nodules and fossils. Some of the stone used for flat cladding panels reveals spectacular coiled ammonite shells—hence the rock's long-winded Italian name.

French Parlor. The fireplace mantel is this site's loveliest example of a heavily veined type of Carrara Marble.

Second-Floor Interior

Radio Room (formerly Charles's Bedroom). This space contains terra-cotta antiquities and a fireplace mantel of chaste white marble that's probably Statuario-grade Carrara. However, when I began writing this book, the real geologic show-stealer here was a magnificently crafted, seventeenth-century stone table. While it's no longer on display, it remains in the museum's collection. For that reason, and because its components are of immense, almost metaphysical significance to the architectural geologist, it's still worth describing here (see figure 8.6).

The table's main ingredient is its speckled, purplish-red Imperial Porphyry. This is arguably the one most famous ornamental rock type in the history of Western civilization. For centuries the ownership of both this luridly gorgeous stone and its quarries was jealously restricted to the emperors of Rome and Constantinople. Any Byzantine ruler who began life in the porphyry-clad birth chamber of the Great Palace rather than attaining the throne as a usurper was dubbed *porphyrogenitus*—"born to the porphyry"—and was assumed to have a greater divine mandate to govern.

Though it has adorned the dwelling spaces of the highest of the high, the Imperial Porphyry comes from one of the most naturally brutal places on Earth, Egypt's searing and sun-scorched Eastern Desert. This barren moonscape is situated between the Nile and the Red Sea. While legend has it that the stone's outcrop area was first discovered by the Roman Caius Cominius Leugas in 18 CE, at least one modern archaeological authority states that it had also been worked much earlier by the ancient Egyptians. At any rate, it was most extensively exploited between the first and fifth centuries CE, when approximately 10,000 tons of rock were removed in giant sections by a multitude of convicts and slaves condemned to this hellhole, or rather hell-hill. The great blocks were laboriously transported down stone chutes to a wadi (dry desert streambed) and then overland to the big river and, ultimately, to the imperial capitals. The Romans knew the stone as *lapis porphyrites* and its mountainous locale, near the modern Jabal Dokhan, as Mons Porphyrites. But after the fifth century all production ceased,

FIGURE 8.6.   A magnificent table made of Imperial Porphyry and the Pietre Dure style of inlaid stonework. While it's no longer on display in the Allis House's Radio Room, this item of supreme geologic interest remains in the museum's collection. (Courtesy of the Charles Allis Art Museum; photograph by Kevin Miyazaki.)

and its site was utterly forgotten. Only in the nineteenth century were the ancient works finally rediscovered by British explorers infused, one suspects, with capitalistic zeal. But they found the place simply too remote and unforgiving to be commercially viable. A single source I've come across states that some new stone recently quarried is now on the market, but even if that's true, its impact is negligible. The vast majority of Imperial Porphyry, including that which makes up this table's top and supporting columns, has been recycled from ancient Roman sites and times. This massive expropriation of the architectural past was done with ruthless aplomb by scalpellini of the Renaissance and Baroque era. They called this rock, once again so highly prized, *porfido rosso Egiziano* or, alternatively, *porfido rosso antico* ("Egyptian red porphyry," "Ancient Red Porphyry"). Interestingly, the stone of this tabletop is the premium brecciated form, the variety most favored by Roman designers and sculptors.

Petrologically, this noblest of rocks is classified rather dauntingly as a trachyandesite-dacite porphyry, which means it's an intermediate to felsic igneous type that has a microcrystalline groundmass peppered with larger phenocrysts. Here the visible crystals are white to pink plagioclase feldspar, sometimes geochemically altered to other compounds instead. But the ruddy matrix itself is a mixture of quartz and both alkali and plagioclase feldspars, all tinted by tiny flakes of hematite. One unit of the Dokhan Volcanic Suite, the Imperial Porphyry was originally considered *extrusive*—lava that reached the surface before solidifying. But a study published in 2018 suggests that instead it's a *shallow intrusive*, which hardened in a magma chamber beneath a volcano. Its age, derived from a remarkably sophisticated technique that analyzes tiny zircons embedded in the rock, ranges from 609 to 600 Ma. That makes it late Neoproterozoic, the time of the Pan-African Orogeny, when two great landmasses merged to form the southern supercontinent of Gondwana.

The table's other geologically derived wonder is its inlaid stone, set in intricate patterns featuring such semiprecious gems as jasper, agate, and lapis lazuli. This roster of resistant minerals and stones indicates that this is a superb example of the Florentine *Pietre Dure* ("hard rocks") method of inlay. A competing technique used by craftsmen in Rome mostly employed Carrara Marble and other softer stones instead. But regardless of the style or exact roster of materials, the craftsmen who made and still make these amazing mosaics must have a superhuman supply of patience, painstaking precision, and first-rate artistic talent.

Sarah's Bedroom. Here there's one more white fireplace mantel, and once again it seems to be Carrara Marble of the pure-white variety.

Sarah's Sitting Room. Our final space and final fireplace. The mantel appears to be made of the brecciated Chiampo Limestone also on display in the Marble Hall downstairs. The museum's records refer to this polishable sedimentary rock as marble, in standard architectural practice.

8.16 Milwaukee-Western Fuel Building

2140–2150 N. Prospect Avenue
Current use: Restaurant and work lofts
Completed in 1934

Architect: Herbert Tullgren
Geologic feature: American Terra Cotta

One of the city's high-water marks of Art Deco terra-cotta art, this relatively modest building actually relies mostly on cream brick for its exterior cladding. Its source is unknown; the last of Milwaukee's own brickyards closed in the 1920s. But the complementing orange, buff, and black terra-cotta is a joy to scrutinize. For more on the geologic origins of this material and the history of its producer, the American works in McHenry County, Illinois, see site 5.21.

The bas-relief spandrels poised between first- and second-story windows should be required viewing for anyone interested in our civilization's Faustian bargain with fossil fuels. The panels portray the coal-mining industry as a mythic enterprise: extracting, transporting, and delivering a cheap and abundant energy source to a frenetic culture arcing ever upward in a state of endless progress. That attitude, that the Earth and her systems would endlessly tolerate our blinkered aspirations, now seems tragically naïve. The fact that we live on a planet with

FIGURE 8.7. At the time the Milwaukee-Western Fuel Building was erected in the 1930s, coal mining—here celebrated on this masterfully crafted American Terra Cotta relief—could still be seen as emblematic of the upward march of civilization.

abundant coal deposits—the remains of Pennsylvanian-subperiod swamp communities and similar environments of other times—has been the best and worst of gifts. After untold centuries of dependence on human- and animal-muscle power, with notable assistance of wind and water, we were suddenly granted accelerated entry into industrialization and all that entails. But the passage of the decades has also revealed the extent and persistence of our addiction, and the mortal danger of wishful thinking and evidence-free optimism. Perhaps someday there will be terra-cotta artwork on other beautiful buildings that illustrates an equally mythic ascent out of our delusions. But that would require our reaching true adulthood.

8.17   North Point Water Tower

2288 N. Lake Drive
Completed in 1873
Architect: Charles A. Gombert
Geologic feature: Wauwatosa Dolostone, Cream City Brick (internal
  structure)

Milwaukee historian John Gurda aptly describes this Victorian Gothic standpipe covering as "a fairy-tale stone tower worthy of the Brothers Grimm." And this storybook structure is the perfect place to study our locally produced, Silurian-period Wauwatosa Dolostone and its weathering characteristics at hand lens range. While this worthy selection was most usually trotted out by Milwaukee architects in its rock-faced form, as it definitely is here, you'll also find it sporting a considerably different look as the tower's dressed-faced trim.

The rock on display here is the decorative covering, so to speak, for the tower's Cream City Brick core. The stone was obtained specifically from the euphoniously named Story Quarry, which at the time was indeed part of Wauwatosa. Later on, however, the quarry locale was ceded to the City of Milwaukee, and it now serves as the northern parking lot of American Family Field.

One minor mystery you'll spot on the tower's exterior is the identity of the igneous inset blocks on each of the four buttresses. These all appear to be the same coarse-grained granitoid rock with hefty feldspar crystals. It's very reminiscent of the gray, coarse-grained variety of Athelstane Granite discussed in the site 5.6 Federal Building section. However, the first company to commercially quarry that stone didn't commence operations till 1888. The insets could have been added later, of course, though they seem to be part of the original design.

FIGURE 8.8. One of Milwaukee's most beloved landmarks, the North Point Water Tower is a Cream City Brick structure clad in Wauwatosa Dolostone quarried in the Menomonee River Valley.

8.18   Oriental Theater

2230 N. Farwell Avenue
Completed in 1927
Architectural firm: Dick & Bauer
Geologic features: Terra-Cotta, Copper

Its maker isn't identified in the sources, but whatever its origin the theater's exterior white terra-cotta is the suitable medium for this Moorish movie-palace style. Also note the copper-clad faux-minaret domes, now nicely weathered with a green, copper-salts patina.

8.19   John F. Kern House

2569 N. Wahl Avenue
Completed in 1899
Architectural firm: Crane & Barkhausen
Geologic features: Jacobsville Sandstone, Terra-Cotta, Wrought Iron

The lovely if unsourced brick and terra-cotta ornament of the exterior are richly complemented with the "Portage Red" brand of Jacobsville Sandstone from Michigan's Keweenaw Peninsula. (See site 5.14 for more on this variety.) It's present in such elements as the plinth, porch arches and columns, balustrades, and window surrounds. The door grille, made of wrought iron heated and then hammered into shape rather than melted and cast, is another nice touch.

8.20   Charles D. Crane House

2603 N. Lake Drive
Completed in 1903
Architect: Charles D. Crane
Geologic feature: Chequamegon Sandstone

In common with the previous site, this residence is German Renaissance Revival in style. But here the rock is Chequamegon Sandstone, specifically taken, the records tell us, from the Prentice Quarry at Houghton Point on Wisconsin's Bayfield Peninsula. It is distinctly more somber in color than the Kern House's "Portage Red." The Chequamegon is more fully described in the St. Paul's Church section (site 8.9).

8.21   George N. Wiswell House

2701 N. Lake Drive
Completed in 1896
Architectural firm: Ferry & Clas
Geologic feature: Chequamegon Sandstone

In my own warped terminology, this delightfully bipolar house is top-grade Tudorranean, with Tudor half-timbered gables, gloomy maroon ashlar, and a jarringly cheerful Mediterranean roof of terra-cotta tiles or some convincing simulacrum. Once again, the rocky portion is the darkling Chequamegon Sandstone from the Prentice Quarry.

The Prentice Brown Stone Company, operating out of North Washburn, Wisconsin, was a major supplier of the Chequamegon to Midwestern markets. As one of its magazine advertisements boasted, it offered its product "in large or small quantities, in blocks for sawing, or other purposes, on board [train] cars or vessel, at short notice." Further, it promised its customers "as good and cheap stone as can be had in this famous Brown Stone District."

8.22   Albert F. Gallun House

3000 E. Newberry Boulevard
Completed in 1914
Architectural firm: Brust & Philipp
Geologic features: Waukesha Area Dolostone, Salem Limestone, Vermont Unfading Green Slate, Wrought Iron

Our foray into the Tudor Revival style continues with this imposing stone-faced mansion. While a lawn-skirting wrought-iron fence protects its residents from roving urban naturalists, rockhounds, and other undesirables, the building's main geologic offerings are on clear display to any underling content to peer at them from the sidewalk.

The house's most abundantly used exterior stone is listed by the architectural historian Russell Zimmermann as "Waukesha County limestone," which could be interpreted as meaning either the Waukesha Area Dolostone, quarried in and about the city of that name, or the famous Lannon Dolostone produced a few miles to the northeast of there. While the Lannon was used extensively as thin-bedded cladding and veneer on twentieth-century Midwestern mansions and is still marketed today, the Gallun House's walls are pure dimension stone, 16 inches thick. This makes the more massive Waukesha Area variety the likelier choice. It's complemented with Salem Limestone serving as trim, carved ornament, and door and window surrounds. The Salem, also known as "Bedford Stone" and "Indiana Limestone," is more fully described at site 5.5.

Another petrological high point of the Gallun House is its thick-tiled roof, identified by Zimmermann as Vermont Unfading Green Slate. This color variant of the bevy of slate selections quarried along the Vermont–New York border owes its tint primarily to the presence of the mineral chlorite. Originally a marine shale deposited in the outer reaches of Laurentia's continental shelf in the Neoproterozoic, it was transformed into foliated metamorphic rock during the succeeding Cambrian and Ordovician periods. This happened when an approaching string of volcanic islands, the Bronson Hill Arc, bulldozed it 60 miles inland onto the continental margin.

8.23   Myron T. MacLaren House

3230 E. Kenwood Boulevard
Other former name: University of Wisconsin–Milwaukee Alumni House
Completed in 1923
Architects: Fitzhugh Scott, MacDonald Mayer
Geologic features: Westwood Granite, Massillon Sandstone, Vermont
   Mixed Slate

Yet another stone-clad Tudor Revival design, the MacLaren House served as UWM's Alumni House for many years, and as such was one of the city's most instructive geologic venues, offering the public on-site parking and a close look at its unusual rock types. Recently, however, university administrators chose to sell the mansion to a local businessman, and it now serves as his private residence. As such, its splendid exterior must be viewed from the distant sidewalk. Even so, its rock selections are so notable they deserve to be described here.

For years, Milwaukee architectural historians have identified the house's polychromatic main cladding as "Plymouth stone, an Eastern quartzite," or words to that effect. Two out of three isn't bad: one of its trade names is indeed "Plymouth Stone," and it does come from the eastern US. But quartzite it most certainly is not. While I'm not sure where that urban legend arose, a happy-go-lucky attitude toward stone identification is so rampant in architectural literature that this particular bit of terminological whimsy requires no further comment. But since this rock isn't really a quartzite (either a very hard, silica-cemented sedimentary sandstone or one that has been metamorphosed), what the blazes is it?

Believe it or not, it's a granite, but everyone who has had trouble identifying it as such should be granted complete amnesty from scorn. As is utterly typical of its variety, it's here so heavily encrusted with weathering products that it's difficult to impossible to discern that it's a fine- to medium-grained intrusive igneous rock composed of a light-tinted mixture of plagioclase feldspar, the alkali feldspar microcline, and quartz. A small amount of black biotite is also present. Officially mapped and described as the Westwood Granite, this rock is

FIGURE 8.9.   Back when the MacLaren House was still a public-access UWM facility, it was possible to examine its cladding stone at close range. But even then it could be difficult to identify. However, on this less weathered section its mixture of black biotite grains and lighter feldspar and quartz crystals indicates that the stone is not the "quartzite" cited by various architectural historians. Instead, it's an igneous intrusive granitoid rock—Westwood Granite from Massachusetts.

late Neoproterozoic in age and quarried to this day in Massachusetts's Plymouth County town of Weymouth. This accounts for its two most common names, "Plymouth Stone" and "Weymouth Granite." But in older sources it's also called "Seam-Faced Granite."

The Westwood began as magma intruded into somewhat older igneous rocks of the Avalonia Terrane, a wayward microcontinent originally associated with Gondwana that collided with what is now New England in the Devonian period. In places in its Weymouth quarrying district it's highly jointed in such a way that at the surface it forms almost vertically oriented sheets from 2 inches to a foot thick. The gaps between the sheets have afforded ample exposure to the elements, and especially the infiltration of water and dissolved minerals. As a result, stone taken from this zone comes ready-coated with deposits of iron oxide and other compounds that give it its palette of brown, silver, ocher, and slate gray. The Westwood is also renowned as an exceptionally hard rock.

Joining the Westwood on the MacLaren House's main elevations is a buff, carved and dressed-faced sedimentary selection that could easily be mistaken for Salem Limestone from just a few paces away. But it's finer in texture and completely nonreactive in dilute hydrochloric acid. Accordingly, a good guess as to its identity would be one of two classic Ohio selections commonly seen in Milwaukee, the Berea Sandstone and the Buena Vista Siltstone. But in fact it's yet a third Buckeye rock type, the Massillon or "Briar Hill" Sandstone. Quarried in the eastern portion of that state, it dates from the Pennsylvanian subperiod, and unlike the Buena Vista and Berea is also available in shades of yellow, pink, and even maroon.

Topping this pair of relatively rare stone types is a roof of varicolored Vermont Mixed Slate. For more on this striking medley of purple, gray-green, and russet metamorphic rock see the Wisconsin Consistory description (site 5.23).

8.24 Milwaukee-Downer College Buildings

E. Hartford Avenue and N. Downer Avenue
Current name: Johnston, Merrill, and Holton Halls, University of
   Wisconsin–Milwaukee
Completed in 1905
Architect: Alexander C. Eschweiler
Geologic features: St. Louis Brick, Jacobsville Sandstone, Terra-Cotta

This ensemble is the most powerful color statement in all of Milwaukee architecture, especially when seen in full sunlight. If City Hall, the Pabst Theater, and the Oneida Street Station are showplaces of pink and orange versions of St. Louis Brick, this is the place to admire its classic bright-red form.

While common brick made for local use in St. Louis yards was fabricated from abundant deposits of Pleistocene loess (windblown silt), the premium facing brick on display here and the sites mentioned above comes from a much more ancient source—the soil in which a complex ecosystem of giant, free-sporing trees and other swamp vegetation sank their roots more than 300 Ma ago. This *paleosol*, which became one of the strata of eastern Missouri's Pennsylvanian-subperiod bedrock, is known as the Cheltenham Clay, and it was mined directly under the city.

The St. Louis brick is trimmed and decorated with red terra-cotta of unknown source, and with a rock type uncited in the sources but almost certainly the "Portage Red" variety of Upper Michigan's Jacobsville Sandstone. This most brightly tinted form of Lake Superior Brownstone is discussed at greater length in the site 5.14 section.

FIGURE 8.10.   The Milwaukee-Downer College buildings, now part of the UWM campus, are a sight to behold on a sunny day, when their bright-red St. Louis Brick exteriors are silhouetted against the blue sky.

8.25   Sabin Hall, UWM

3413 N. Downer Avenue
Completed in 1928
Architectural firm: Van Ryn & DeGelleke
Geologic features: Temiscouata Slate, Terra-Cotta

The Wisconsin Historical Society description of this building states that its exterior is made of deep-red brick, origin not cited, and an unspecified sandstone. The brick is obviously here, but most or all of what seems to be the "Portage Red" variety of Jacobsville Sandstone is in fact an unsourced terra-cotta made to simulate that rock type's unique tint. This certainly wouldn't be the first time that mimetic material was mistaken for building stone. And, as a matter of fact, by the time Sabin Hall was constructed, the Jacobsville was no longer being quarried for architectural use. Additionally, there are at least a couple of places by the entrance where the trim has been chipped to reveal pale-yellow *bisque*, the fired-clay base of terra-cotta on which the glaze is applied. These dents show that the "Portage Red" color is just a coating less than a single millimeter thick.

The item of greater geologic interest is the roof tile, which, depending on the lighting, looks silvery, dull gray, or somewhat darker. This is the "North Country Unfading Black" brand of the Temiscouata Slate. The other sites in this book that feature this foliated, fissile, and remarkably useful rock type offer examples of varieties produced in Vermont, New York, and Pennsylvania. But the Temiscouata is Canadian and comes specifically from the Glendyne Quarry, in Saint-Marc-du-Lac-Long, Quebec. A glance at this location on the map will convince you that it's the very essence of the eastern Quebecois wilderness: glacially abraded terrain now covered with birch-and-conifer forest. The rock extracted here is lower Devonian and began as mud of *turbidite* (marine avalanche) deposits that accumulated in the foreland basin of the Acadian Mountains. Continued crustal compression ensured its transformation from sedimentary shale to metamorphic slate. "North Country Unfading Black" remains one of North America's most reliable, enduring, and popular of roofing slates, and for good reason it's used extensively in the US.

8.26   Alfred M. Hoelz House

3449–3451 Frederick Avenue
Completed in 1925
Architectural firm: Arnold F. Meyer & Company
Geologic feature: Crossville Sandstone

Like Tosa's geologically notable Fiebing House, the exterior of Downer Woods's Hoelz House is decked out in Crossville Sandstone ("Tennessee Quartzite") from the Cumberland Plateau. Here, however, the multicolored flags are set not in rubble fashion but in something approaching regular courses. For more on the Crossville, see site 6.34.

8.27   St. Robert Catholic Church

2214 E. Capitol Drive, Shorewood
Completed in 1937
Architectural firms: Maginnis & Walsh; L. A. Brielmaier & Sons
Geologic features: Milbank Area Granite, Salem Limestone, Terra-
    Cotta Roof Tile

A re-creation of the Romanesque-style variant characteristic of northern Italy's Lombardy region, St. Robert Church definitely has a Mediterranean look to it, thanks to its roof of terra-cotta tile. Equally striking are the main portions of the elevations and bell tower, which are composed of a colorful medley of bricks ranging from palest pink to russet.

The façade also features two stone types well known to twentieth-century American architects. The ever-obliging Salem ("Bedford," "Indiana") Limestone is present in force as hefty sections of trim and all sorts of sculpted detail, including the Corinthian capitals of the two entrance columns. The shafts under them, however, are something completely different—polished Milbank Area Granite. It's probably the "Dakota Mahogany" variety quarried in Milbank, South Dakota. By no means as old as Minnesota's Morton Gneiss, it's extremely venerable nonetheless. Upper Neoarchean in age, it has been dated to approximately 2.6–2.7 Ga. It comes from the Benson Block, a section of its state's Precambrian basement that is part of the Minnesota River Valley Terrane. That means that it hails from one of our continent's ancient cores, the Superior Craton.

The Milbank Area Granite, still actively produced today, is a common sight in Chicago and its suburbs, and is probably found throughout this region, too. That said, the only other documentation I've seen so far for its presence here is that for Milwaukee's now-demolished Bradley Center. It had exterior cladding of the "Carnelian" variety (see figure 1.1). In any case, the Milbank is an overtly gorgeous, coarse-grained rock, with large crystals of pink orthoclase, gray quartz, and black biotite. The fact that it lacks a significant complement of plagioclase feldspar as well indicates that it's what petrologists call an *alkali-feldspar granite*. This type includes an assortment of other red granites used architecturally.

8.28 Radio City

720 E. Capitol Drive
Alternative name: WTMJ Studios
Completed in 1941
Architectural firm: Eschweiler & Eschweiler
Geologic features: Oneota Dolostone, Sgraffito

While the majority of Milwaukee's most significant expressions of the Art Moderne style are clothed in gray-to-buff Salem Limestone, this sleek expression of rounded elegance in the horizontal plane has instead an exterior of the "Kasota Stone" variant of Minnesota's Oneota Dolostone. As usual, this Ordovician carbonate sedimentary rock presents a subtly complex palette of yellow, pink, and cream tones. See sites 6.19 and 6.25 for more on the origins of this popular selection.

For eight decades, Radio City was home to WTMJ-AM. It remained here, along with three sister radio stations, until 2021, when all were moved downtown by their current owner. However, the building's largest occupant, WTMJ TV (Channel 4), is still present as this book goes to press. If possible, go in the main entrance and take a good look at the lobby's four walls. They're all decorated with unusual *sgraffito* tableaux created by artist Jefferson Greer in 1941. Greer applied brightly tinted layers of that eminently Earth-derived material, plaster, and then skillfully scraped away portions of it to reveal underlying colors in just the right places. In doing so he produced a sequence of murals that include female figures symbolizing various aspects of broadcasting. They loom over radio-studio musicians, television-program performers, and rapt urban and rural listeners. Here Greer cleverly adapted a time-honored decorative technique especially popular in the Renaissance to the stripped-down, streamlined design ethic of his own time. From a geologic perspective the result is a most interesting anthropogenic deposit of gypsum or lime mixed with sand, water, and coloring agents.

8.29   Estabrook Park

4400 N. Estabrook Drive, Shorewood
Geologic features: Milwaukee Formation Dolostone, former Milwaukee
   Cement Company quarrying locale
Note: Please do not collect fossils or damage rock outcrops

No exploration of Milwaukee County's geology would be complete without a visit to the falls site adjacent to Estabrook Park's culturally appropriate Beer Garden. For here we see our *other* bedrock, nicely shorn of its Quaternary-sediment burden by downcutting of the Milwaukee River. We've moved one rung up in the periods of the geologic time scale, from the Silurian strata that in Wauwatosa and elsewhere provided excellent dimension stone. As it turns out, the Middle Devonian rock on view here also played a crucial role in the architectural development of this region and, for that matter, much of the nation. This because it yielded a much sought-after building material—not blocks of ashlar, but the raw material for Milwaukee's once-thriving cement industry.

The low falls here certainly appear to be a pleasant bit of natural scenery, especially when adorned with wading weekend fishermen angling for sturgeon, coho salmon, and rainbow trout. But in fact the falls are a manmade feature that was formerly the northern wall of an old Milwaukee Cement Company quarry. That quarry was active in the days when the river's flow was diverted to the east of here. In this area the lowest, Berthelet Member of the Milwaukee Formation

Dolostone was extracted in vast quantities because its mineral composition and relatively high clay content made it perfect for the production of *natural hydraulic cement*. Unlike the Portland cement that ultimately outcompeted it, the natural type owed its quality and remarkable ability to set even underwater to its source rock's own intrinsic composition. No other ingredients needed to be shipped in from elsewhere. In contrast, the Portland variety required the addition of separately obtained clay, and it probably would not have prevailed ultimately had not improvements in lime-kiln technology and lowered transportation costs made it more commercially viable. At any rate, the Milwaukee Cement Company, which began quarrying and production here in 1876, was forced to cease operations by 1911. And its local competitors fared no better.

Nowadays in this peaceful, leafy setting there's little sign of the massive excavations and industrial landscape that once were here. But if you stroll downstream from the falls along the eastern-bank path, you can still see exposed Milwaukee Formation Dolostone in various spots. This gray, argillaceous carbonate rock dates to about 390 Ma and formed in typical sea-bottom conditions. As local geologists and paleontologists are well aware, this rock unit is very fossiliferous and justly noted for the diversity of its identified species. It features not only a rich assemblage of such marine invertebrates as trilobites, cephalopods, bryozoans, and brachiopods; it also contains the remains of fish, some of the earliest trees, and *Prototaxites milwaukeensis*, a monstrous, upright fungus that could grow to 20 feet in height. And, as you may discover yourself, it has vugs often filled with *bitumen*. This tarry compound is the residue of ancient organisms that seeped into underlying strata.

For more on both the amazing paleontological legacy of this site and its history of cement production, see the highly recommended *Fossils of the Milwaukee Formation*, by Gass, Kluessendorf, Mikulic, and Brett. It's listed in the Selected Bibliography.

Glossary

ABBREVIATIONS

cm = centimeters
Ga = billion years
in = inches
ka = thousand years
km = kilometers
Ma = million years
mi = miles
mm = millimeters

Underlined items in the text have their own glossary entries.

Acadian Orogeny The orogeny that occurred in the Devonian period when Avalonia and other crustal fragments collided with Laurentia.

Acroterium (also rendered Acroterion) A vertically projecting ornamental element placed atop or on the edges of a pediment. The plural form of both singular spellings is acroteria.

Aggregate Particulate material such as sand, gravel, or rock fragments that is mixed with cement to produce concrete.

Alkali Feldspars One of the two main groups of feldspar minerals. Its members are the potassium feldspars (orthoclase, microcline, and sanidine) and anorthoclase.

Alkali-Feldspar Granite A granitoid rock that differs from regular granite in that at least 90 percent of its feldspar content is in the form of alkali feldspars.

Amazonia An ancient continent that now forms a significant portion of South America.

Ammonite A type of cephalopod especially prevalent in the Mesozoic era.

Amphibolite A green to black, weakly foliated metamorphic rock that has a variety of parent rocks, including basalt and komatiite.

Aniconic Referring to artwork or designs that contain no human or divine imagery.

Anorthoclase A silicate mineral of the alkali-feldspar group. It sometimes exhibits the properties of labradorescence.

Anorthosite A felsic igneous rock that is composed almost completely of plagioclase feldspar but may also contain in some cases significant amount of pyroxene. It often exhibits labradorescence.

Anthropogenic Produced by human activity; manmade.

Aphanitic Composed of microscopic crystals.

Aplite A granite of unusually fine-grained texture.

Apse The end section of a church interior, often semicircular in plan, that houses the altar.

Arenaceous Containing sand particles, or sandy in texture.

Archean Eon The span of geologic time from 4.0 to 2.5 Ga.

Archivolt The ornamental molding or surface on the underside of an arch.

Argillaceous Containing clay particles.

Art Deco A term, derived from the Exposition des Arts Décoratifs et Industriels Modernes, for the architectural and decorative style prevalent especially in the 1920s. Its salient features include an emphasis on elegant, stylized ornamentation and streamlined and rectilinear forms inspired by machines and technology rather than by natural or organic forms.

Arthropod An invertebrate animal of the phylum Arthropoda. Arthropod types include insects, spiders, crustaceans, and trilobites.

Art Moderne The term referring to the leaner, less highly ornamented 1930s offshoot of Art Deco.

Ash Tephra that has a particle diameter of 2 mm or less.

Ashlar Quarried stone that has been "squared off " (cut with sides that meet at right angles). Ashlar blocks may be arranged in regular courses or in a more random pattern.

Avalonia The microcontinent that apparently formed on the margin of Gondwana but later broke off and migrated to collide with both Baltica and Laurentia, where it triggered the Acadian Orogeny.

Baltica The ancestral version of northern Europe that existed in Proterozoic and Paleozoic time.

Baluster A small and often decorative column or pillar that supports either a horizontal or staircase handrail.

Balustrade A series of balusters surmounted by a handrail.

Banded Iron Formation A chemically precipitated sedimentary rock, most usually dating to the Archean or Paleoproterozoic, composed of alternating bands of chert and hematite, magnetite, or other iron-containing minerals.

Barrel Vault A ceiling of half-cylindrical form.

Basalt An aphanitic, quartz-poor, and usually dark-colored igneous rock. It is the extrusive equivalent of gabbro.

Bascule A counterweighted architectural element, such as a bridge span, that rotates around a horizontal hinge line.

Base Course The lowest course of cladding or masonry on a wall.

Bas-Relief A figure or image that projects in shallow relief from a wall or other surface.

Batholith A large pluton at least 40 square mi (100 square km) in area at the surface.

Bauxite A sedimentary rock that often forms from tropical soils or the weathering of carbonate strata. Of great economic significance, it is the primary ore of aluminum.

Bay An architectural element that projects outward from the main wall.

Beaux Arts Style An architectural style incorporating such classical elements as columns, pilasters, arches, domes, balustrades, medallions, and other Greek- and Roman-derived ornament to produce a sense of grandeur, symmetry, and formalism.

Belt Course A continuous course of brick, stone, or terra-cotta that provides decorative contrast.

Bedded Layered.

Bedding A synonym of strata.

Bedford Tree A synonym of rustic monument.

Bedrock Any section of rock that is still attached to the crust.

Benthic Referring to bottom-dwelling marine organisms.

Biocalcarenite A type of limestone chiefly composed of very small shell fragments and tiny whole fossils.

Biotite A black or dark-brown, sheet-forming silicate mineral of the mica family. One of its listed chemical formulas is $K(Mg, Fe)_3(AlSi_3O_{10})$; other variations are cited.

Bioturbation The disturbance or reworking of sediments by organisms.

Bisque In the terra-cotta trade, the fired clay that serves as the base for the glaze, which is subsequently fixed with a second firing.

Bitumen A black, tarlike substance made of hydrocarbons and derived from organic remains.

Bivalve The type of mollusk that possesses a pair of hinged valves. Examples include clams, oysters, and mussels.

Black Granite A term, most usually an oxymoron, used widely in the quarrying and architectural trades. True granite by its mineralogical composition cannot be thoroughly black, though on rare occasions it can be rather dark-toned. This misnomer is used for such igneous rock types as gabbro, diabase, basalt, and some forms of anorthosite.

Book Match The symmetrical pattern created when two panels of cladding with identical patterns (usually cut in the quarry from the same block of stone) are mounted on a wall side by side, with one panel facing in a direction opposite the other's. This creates an effect of an opened book with two mirror-image pages visible.

Boulder A detached rock larger than 256 mm (10 in) in diameter.

Bowen's Reaction Series A description formulated by geologist Norman Bowen that shows the sequence in which silicate minerals crystallize as a magma cools.

Brachiopod A two-valved marine animal of the phylum Brachiopoda.

Breccia A clastic rock composed of coarse, angular rock fragments.

Brecciated Referring to a marble or other stone type that features large angular or pointed rock fragments separated by an extensive network of veins.

Brick A shaped, rectangular unit of building material made of fired clay (which is sometimes combined with other substances).

Brise Soleil An architectural element that provides partial shade to a building, often using a series of slats, baffles, or parallel ribs.

Bronson Hill Arc A volcanic island arc that collided with Laurentia in the Ordovician period. This terrane now forms western New Hampshire, eastern Vermont, and portions of other New England states.

Brownstone An architectural term for a red-, brown-, or maroon-tinted sandstone. This type of dimension stone was particularly popular with architects who adhered to the late nineteenth-century Richardsonian Romanesque style.

Brutalist Referring to the modernist architectural style that emphasizes simplicity of shape and bare surfaces over grace, intricate ornament, or symmetry.

Bryozoan Referring to colony-forming marine animals of the phylum Bryozoa.

Bush-Hammered Finish A finish created by treating a stone surface with a multipointed hammering tool. This produces a rugged, nonskid surface.

Caisson A vertical shaft, well, or pier dug under ground and filled with concrete or other hard material. This shaft serves as an anchor that connects a building to the underlying hardpan or to the bedrock beneath it.

Calcareous Containing calcium carbonate.

Calciphilic Referring to organisms that thrive on limestone or other calcium-rich environments.

Calcite The carbonate mineral composed of calcium carbonate ($CaCO_3$).

Cambrian Period The span of geologic time from 541 to 485 Ma.

Capital The uppermost section of a column.

Carbonate Mineral A member of a chemical group of minerals that contain, among other substances, CO_3.

Carbonate Platform A marine shelf mantled in carbonate sediments. Carbonate platforms exist only in tropical or subtropical waters; one modern example is the Bermuda Banks.

Carbonate Rock A rock primarily composed of such carbonate minerals as calcite (calcium carbonate) or dolomite (calcium-magnesium carbonate).

Carboniferous Period The span of geologic time from 359 to 299 Ma.

Case Hardening The weathering process by which a rock's outer portion becomes harder and shell-like, while its interior turns soft and crumbly. This is often due to the outward migration of mineral-laden water previously absorbed by the rock.

Cement A liquid substance used in the production of concrete that, when mixed with aggregate or other materials, sets and binds them together. Modern cement is usually composed of lime, gypsum, and clay.

Cenotaph A monument commemorating a person whose remains are lost or buried elsewhere.

Cephalopod A marine mollusk of the class Cephalopoda. Members of this class include the nautiloids, octopi, squid, and the now-extinct ammonites.

Chalcocite A sulfide mineral with the formula Cu_2S. It is a major source of commercial copper.

Chalcopyrite A sulfide mineral with the formula $CuFeS_2$. It is a major source of commercial copper.

Chamfer The beveled or recessed edge of a block of stone.

Chemically Precipitated Rock Referring to a sedimentary rock type that is formed by the precipitation of mineral crystals from water.

Chert A very hard, chemically precipitated sedimentary rock composed of microcrystalline quartz.

Chlorite A term referring to a group of often greenish silicate minerals associated with low-grade metamorphic rocks.

Chromite An oxide mineral with the chemical formula $FeCr_2O_4$. It is the ore of the metallic element chromium.

Cinerary Urn A closed container holding the cremated ashes of a deceased person.

Cipollino An Italian term that literally means "resembling chives or baby onions." It refers to marble that has a banded or thinly layered appearance.

Cladding Stone used on the exterior of a building for decorative effect. Usually it is not a load-bearing element.

Classic Milwaukee Formula The author's term for the use of locally produced materials on Milwaukee building exteriors. The Formula comes in three versions: (1) Wauwatosa Dolostone only; (2) Cream City Brick only; and (3) Cream City Brick above a Wauwatosa Dolostone base.

Clast A rock particle or rock fragment.

Clastic Rock Referring to a sedimentary rock type that is formed of cemented particles—boulders, cobbles, pebbles, sand, silt, or clay—originally produced by the weathering or erosion of other, older rock.

Clay A mineral particle less than .003 mm (.0001 in) in diameter.

Coarse-Grained Referring to igneous or metamorphic stone types whose feldspar crystals are greater than 1.0 cm in length.

Cobble A rock particle 64–256 mm (2.5–10 in) in diameter.

Coffer A ceiling section that is progressively recessed toward its center. A pattern of coffers, such as that on the dome of Rome's Pantheon, can create a striking visual effect.

Cofferdam A enclosure erected to keep water out of an excavated area.

Column A pillar, usually circular in cross section, that serves as either a load-bearing or merely ornamental element.

Common Brick Brick, usually of lesser ornamental appeal and made of unscreened or less carefully selected clay, that is used for building sides and rears, and for interior walls not intended for display. It is often softer, less sharp-edged, and less regular in shape than face brick.

Composite Order The classical architectural order that has capitals combining the volutes of the Ionic order with the acanthus leaves of the Corinthian order.

Concrete A thick liquid mixture of <u>cement</u>, <u>aggregate</u>, water, and other substances that sets into a hard, durable, solid material widely used for construction.

Conglomerate A <u>clastic</u> <u>sedimentary</u> <u>rock</u> composed of <u>coarse</u>, rounded fragments.

Contact Metamorphism Localized <u>metamorphism</u> of preexisting <u>rock</u> caused by the intrusion of a nearby body of <u>magma</u> or <u>hydrothermal fluids</u>.

Continental Crust The part of the Earth's <u>crust</u> that forms the continents. It is largely composed of <u>felsic rock</u> types.

Continental Shelf An extension of a continental margin covered by relatively shallow saltwater.

Convergent Plate Boundary A zone where two <u>plates</u> move toward each other.

Cordate Heart-shaped.

Core The center zone of the Earth's interior, about 3,500 km (2,200 mi) in diameter, just below the <u>mantle</u>. It is composed of a liquid iron-nickel outer core and a solid iron-nickel inner core.

Corinthian Order The classical architectural order that features relatively slender <u>columns</u> with fluted shafts and ornate <u>capitals</u> decorated with acanthus leaves.

Corundum An extremely hard <u>oxide mineral</u> with the chemical formula Al_2O_3. The precious stones ruby and emerald are forms of corundum.

Course A horizontal row of <u>bricks</u> or <u>dimension stone</u>.

Craton The <u>tectonically</u> stable interior of a continent.

Crazing A network of fine cracks that develops on the surface of a glazed, burnt-<u>clay</u> product.

Cretaceous Period The span of geologic time from 145 to 66 Ma.

Crinoid A marine animal, often <u>sessile</u> in its adult form, commonly found in <u>Paleozoic</u> <u>fossiliferous strata</u>. Popularly referred to as "sea lilies," crinoids are members of the phylum Echinodermata.

Crossbedding A pattern of narrowly spaced and slanting or curving layers found within the <u>strata</u> of <u>sandstone</u> and other <u>sedimentary</u> <u>rocks</u>. Crossbedding indicates the general direction and force of the wind or water that originally laid down the <u>sediments</u>.

Crust The uppermost section of the solid Earth, ranging in thickness from about 6 to 50 km (4 to 30 mi). It lies directly above the <u>mantle</u> and below the atmosphere.

Crustal Referring to the Earth's <u>crust</u>.

Crystalline Rock A general term for any <u>igneous</u> <u>rock</u> or <u>metamorphic</u> <u>rock</u> of <u>igneous</u> parentage.

Cumulate A term referring to an <u>igneous rock</u> formed of crystals that either sank to the bottom or floated to the top of a body of otherwise still-molten <u>magma</u>.

Dacite An <u>extrusive</u>, <u>intermediate</u> to <u>felsic</u> <u>igneous</u> <u>rock</u>.

Damp Course A <u>base course</u> made of nonabsorbent <u>stone</u> or another material that resists dampness and the uptake of ground moisture.

Devonian Period The span of geologic time from 419 to 359 Ma.

Dike A narrow, vertical or steeply angled, wall- or stripelike mass of <u>intrusive</u> <u>rock</u> that forms when upward-moving <u>magma</u> cuts across preexisting <u>rock</u> bodies.

Dimension Stone Any quarried <u>rock</u> product that is cut to a specific size or shape for architectural or construction uses.

Distributary One of the branching channels of a stream flowing across a delta.

Divergent Plate Boundary A zone, characterized by rifts and volcanic activity, where the older portions of two <u>plates</u> are moving away from one another.

Dolomite The <u>carbonate mineral</u> $CaMg(CO_3)_2$. Also, a term used to denote the <u>rock</u> type chiefly made up of this <u>mineral</u>.

Dolostone The less ambiguous term for the <u>rock</u> type chiefly composed of the <u>mineral</u> <u>dolomite</u>.

Doric Order The classical architectural order that features columns with fluted shafts and simple, round capitals.

Dormer A window that projects upward and outward from a building's roof.

Dressed Referring to stone that has been prepared for architectural use. This preparation includes creating an essentially plane exposed face and also often applying an ornamental or protective finish to the exposed face. The finish can be applied with mechanical tools, abrasives, or chemical treatments.

Drift A collective term for material deposited directly or indirectly by a glacier.

Dunite A type of peridotite with a composition that is at least 90 percent olivine. It forms in the upper mantle beneath oceanic crust.

East-Central Minnesota Batholith A complex of Paleoproterozoic plutons dating to approximately 1.78 Ga that underlie Minnesota's Mississippi and Sauk River Valleys in the vicinity of the town of St. Cloud.

Eastern Granite-Rhyolite Province A terrane composed of largely felsic igneous rock, 1.55 to 1.35 Ga old, that accreted to Laurentia in the region that is now the eastern and central portions of the American Midwest.

ECMB The abbreviation for East-Central Minnesota Batholith.

Efflorescence A white weathering crust produced when salt solutions migrate through porous stone and brick and precipitate on the surface.

Emery An abrasive mixture containing corundum and either hematite or magnetite.

Enamel A coating, usually containing silica, which fuses at very high temperature directly with fired clay or metal to produce a sealant coating more enduring and craze-resistant than a glaze.

Enameled Steel Steel that has fused with a vitreous enamel to produce an ornamental, easily cleaned, and corrosion-free architectural element.

End Moraine A ridge of till deposited at the leading margin of a glacier when the rate of ice melting at the margin is matched by new ice moving up from the rear.

Eoarchean Era The span of geologic time from 4.0 to 3.6 Ga.

Eocene Epoch The span of geologic time from 56 to 34 Ma.

Eon The largest subdivision of the geologic time scale.

Epeiric Sea An extensive but relatively shallow body of saltwater that covers a portion of a continent during a time of high global sea level.

Epicontinental Referring to a sea that covers a portion of a continent.

Epoch On the geologic time scale, the largest subdivision of a period.

Era On the geologic time scale, the largest subdivision of an eon.

Erosion The process by which sediments and other Earth materials are removed by wind, running water, or glaciers.

Erratic A detached rock that had been transported by a glacier and subsequently deposited on the ground when the glacier melts.

Exfoliation *See* Scaling.

Extrusive Referring to igneous rock formed by the cooling of lava or tephra on the Earth's surface (either on dry land or on the floor of a body of water).

Façade The front-facing portion of a building's exterior.

Face Brick Brick of higher quality, hardness, durability, as well as greater ornamental appeal, that is used for building façades and interior walls intended for display.

Faience A term for glazed ceramic tile used for architectural ornament.

Fault A fracture in the crust where there has been significant displacement between the two sides.

Federal Style A classically derived architectural style frequently employed in American houses of the late eighteenth and early nineteenth century. Federal buildings often have flat external surfaces with sparing ornament.

Feldspar A term for any member of the very prevalent complex of silicate minerals that contain, among various other elements, aluminum, silicon, and oxygen.

Felsic Referring to igneous rocks rich in silicon and aluminum. Such rocks include granite, syenite, tonalite, quartz monzonite, and rhyolite.

Ferromagnesian Mineral A silicate mineral, usually dark-colored, that contains a significant amount of iron and magnesium.

Fieldstone (Architecture) Referring to an architectural style that features uncut fieldstones set into the façade of a building.

Fieldstone (Geology) A detached cobble or boulder (in our area, often an erratic) that is collected from farm fields and other open ground for architectural and construction purposes.

Fine-Grained Referring to igneous or metamorphic stone types whose feldspar crystals are less than 0.5 cm (0.2 in) in length.

Finish The texture and surface appearance of a stone after its treatment by mechanical or chemical means. A finish is applied to either enhance a stone's aesthetic appeal or to better preserve it from weathering and wear.

Fissile Referring to a rock that can be readily split into sheetlike or slablike sections.

Flagging The setting of flagstones to create a durable, flat surface.

Flagstone A flat, thin-bedded stone of sufficient hardness and durability to be used as a path, sidewalk, patio, or roadway surface.

Flamed Finish Synonymous with Thermal Finish.

Flemish Bond A type of brickwork that consists of courses of alternating headers and stretchers.

Flow An outpouring of lava on the Earth's surface.

Fluvial Clay Clay deposited by stream flow.

Flysch A sequence of deep-water sedimentary rocks (especially shales and turbidites) that accumulate in a foreland basin.

Foliated A term applied to metamorphic rocks that have minerals in a parallel alignment that often gives them a banded, wavy, or thinly layered appearance.

Foraminifer or Foram A type of planktonic marine organism, usually microscopic or almost so, that grows a calcareous test ("shell"). The plural form is **Foraminifera**.

Forearc The region between a subduction zone and a volcanic arc.

Foreland Basin A basin, often filled with seawater, that forms adjacent to a mountain range.

Fossil A preserved part or indication of an ancient organism's form or behavior.

Fossiliferous Containing fossils.

Foundation The portion of a building, situated either completely or mostly underground, that anchors the rest of the structure in the substrate.

Freestone A quarryman's term for a sedimentary rock that can be easily sawn or cut in any direction without unwanted splitting, and thus is easily worked.

Frit A powder made of a ground glass that is fused with metal at high temperature to produce a vitreous enamel coating.

Fucoidal An old term used to describe the texture of rock containing tubelike and branched trace fossils, apparently burrows of benthic marine organisms of the genus *Thalassinoides*.

Gabbro A phaneritic, dark-colored, and quartz-poor igneous rock. It is the intrusive equivalent of basalt.

Gable The triangular portion of a building between two slopes of the roof that meet at their top.

Ganderia A microcontinent that collided with Laurentia in the interval between the Taconic Orogeny and the Acadian Orogeny. It has been hypothesized that Ganderia originally broke off the margin of Gondwana.

Gastropod An invertebrate mollusk of the group that includes snails.

Geologic Time Scale The visual representation, often drawn to scale and displayed as a horizontal or vertical bar, of the Earth's history, from its origin approximately 4.54 Ga to the present. It is divided into these subdivisions of decreasing order and size: eons, eras, periods, and epochs.

Geothermal Referring to hot water or hot-water solutions.

Glaciation An episode (often lasting about 100 ka) consisting of the formation, advance, and retreat of one continental ice sheet. Not synonymous with ice age (which see).

Glacier A large, persistent mass of ice that moves outward under the weight of its own thick center on low slopes, or downward on steep valley slopes. A glacier forms when there is, for an extended period, more net annual snowfall than net annual melting.

Glaze A compound applied to a fired clay as a sealant or for ornamental effect. Unlike enamels, glaze compounds are not heated to a temperature high enough to cause direct fusing with the underlying material.

Gneiss A high-grade metamorphic rock characterized by linear bands of dark and light minerals.

Gneissic A term applied to granites and other similar igneous rocks that show some evidence of banding or foliation suggestive of some degree of metamorphism.

Goethite An oxide mineral of the chemical formula FeO(OH).

Gondwana or Gondwanaland The Neoproterozoic- and Paleozoic-era supercontinent that included Africa, India, South America, Australia, and Antarctica. The term is also used for the southern portion of Pangaea.

Gothic Revival Style A largely nineteenth-century architectural style that emulated medieval Gothic designs typified by such features as pointed arches, buttresses, and rib-vaulted ceilings.

Graben A down-dropped crustal block.

Grade The current surface level; ground level.

Grade Course A course of brick or stone, often the bottommost on a building exterior, at grade.

Grainstone A type of carbonate rock in the Dunham classification system that is composed of grains not surrounded by a mud matrix.

Grand Art Deco Formula A term coined by the author to describe a common design pattern found in skyscrapers and other buildings embodying the Art Deco and Art Moderne styles. It consists of a massive, uniform Salem Limestone exterior contrasted at its base by at least somewhat darker, crystalline rock.

Granite (Architecture) An imprecise and variably defined term that generally refers to any hard, silicate rock type (be it igneous, sedimentary, or metamorphic) that can take a high polish.

Granite (Geology) A phaneritic, usually light-colored, felsic, igneous rock. The intrusive equivalent of rhyolite, it contains at least 10 percent quartz.

Granite-Rhyolite Province *See* Eastern Granite-Rhyolite Province

Granitoid Referring to the group of igneous intrusive rocks that includes granite and such closely related types as alkali-feldspar granite, granodiorite, quartz monzonite, aplite, and tonalite.

Granodiorite A felsic to intermediate granitoid rock that has 65 to 95 percent of its feldspar content in the form of plagioclase, and is 20 to 60 percent quartz.

Gravel A general term for rock particles of pebble size.

Grillage A type of shallow building foundation made of a network of crisscrossed timber beams or iron or steel rails.

Groundmass The matrix of small crystals in an igneous rock in which phenocrysts are set.

Groundwater Water held beneath the Earth's surface in rock, sediments, or the soil.

Gypsum The sulfate mineral $CaSO_4 \cdot 2H_2O$.

Hadean Eon The span of geologic time from 4.54 to 4.0 Ga.

Halide Mineral A member of a chemical group of minerals that contain, among other substances, a halide element (fluorine, chlorine, bromine, etc.).

Halite The halide mineral NaCl. In its rock form it is also known as "rock salt."

Hardpan A zone of soil or unconsolidated sediments that has been compacted into a stiffer, almost rocklike consistency.

Hardpan Caisson A deep caisson that extends downward to the hardpan layer above the bedrock.

Header A brick or block of dimension stone in ashlar masonry that is set with its short face exposed.

Headwall The vertical working face of a quarry from which stone is extracted.

Hematite The oxide mineral Fe_2O_3, a principal iron ore.

High-Grade Metamorphic Rock A metamorphic rock that has been subjected to a great increase in pressure, and to temperatures above 320 degrees Celsius (608 degrees Fahrenheit).

Highstand A time of high sea level.

Honed Finish A finish created by treating a stone surface with abrasives in a way that results in a nonreflective surface and somewhat duller colors.

Hornblende A common, black, rock-forming silicate mineral with the general formula of $(Ca,Na)_{2-3}(Mg,Fe,Al)_5(Si,Al)_8O_{22}(OH,F)_2$.

Hot Spot The surface expression of a mantle plume. Hots spots are characterized by dramatic volcanic activity and massive lava flows.

Hydraulic Cement Cement that hardens underwater and in the absence of free oxygen.

Hydrothermal Fluids Hot and chemically reactive fluids circulating underground.

Ice Age A major, extended event in Earth history (often lasting 10 Ma or more) that includes at least several glaciations and interglacials.

Igneous Referring to any rock formed by the cooling of magma, either underground or at the Earth's surface.

Intermediate Referring to igneous rocks that have a composition midway between that of felsic and mafic types. Such rocks include monzonite and some granodiorites.

Interreef Referring to rock strata that form from sediments that are deposited between or at some distance from reefs.

Intrusive Referring to igneous rock formed from magma cooling beneath the Earth's surface.

Inverted-Arch Foundation A foundation that supports a building's load-bearing walls with piers connected to arches set upside down.

Ionic Order The classical architectural order that features columns that are usually fluted and have capitals with volutes.

Island Arc A curving line of volcanic islands that forms over a subducting plate.

Isotope A form of an element that has atoms with a nonstandard number of neutrons.

Isotopic Age *See* Radiometric Age.

Joint A fracture in bedrock where there has been no significant displacement between the two sides.

Jurassic Period The span of geologic time from 201 to 145 Ma.

Kame A conical, ridgelike, or terracelike landform made of sand, gravel, or cobbles deposited in a depression or cavity in a stagnant glacier, or at the base of a waterfall on the margin of a stagnant glacier.

Karst A terrain of carbonate rocks characterized by such solution features as sinkholes, caves, and disappearing streams.

Komatiite A rare extrusive igneous rock that is the ultramafic equivalent of basalt.

Knot A quarryman's term for a clump or clot of dark minerals in a granitoid rock that is considerably larger than the surrounding crystals. Knots were often considered unsightly detractions from a stone's appearance, but one can find them nevertheless on display in buildings and monuments made of such <u>rock</u> types as Minnesota's Hinsdale Granite, where they can be huge, and in Nova Scotia's Nictaux Granodiorite.

Labradorescence A property of some <u>minerals</u> that have an internal structure composed of a plane of molecules that reflects light to produce iridescent colors. Also referred to as schiller.

Lacustrine Deposit A <u>sediment</u> deposit laid down on the bottom of Lake Michigan or another lake.

Lagerstätte A <u>sedimentary</u> <u>rock</u> unit much prized by paleontologists because it contains well-preserved <u>fossils</u> of soft-bodied creatures not generally found elsewhere.

Lagging As used in architecture and engineering, wood slats used to line, stabilize, and seal the sides of <u>caissons</u> before they are filled with <u>concrete</u>.

Landform Any distinct feature on a planet's surface produced by natural forces.

Laurasia The northern section of the <u>supercontinent</u> of <u>Pangaea</u>.

Laurentia The ancestral version of North America as it existed in the <u>Proterozoic eon</u> and early <u>Paleozoic era</u>.

Laurussia The continent that formed in the <u>Devonian period</u> from the collision of <u>Laurentia</u>, <u>Avalonia</u>, and <u>Baltica</u>.

Lava Molten <u>rock</u> at the Earth's surface (either on land or on the ocean bottom).

Leucogranite A <u>granite</u> that is light-colored owing to an absence of substantial dark-hued <u>mineral</u> content. Leucogranites are thought to be the product of continental collision; their <u>magmas</u> are derived from <u>metamorphosed</u> <u>sedimentary</u> <u>rocks</u> found in the upper <u>crust</u>.

Liesegang Ring A linear pattern, often curving, wavy, or ringlike, produced by iron-oxide <u>minerals</u> precipitating out of <u>groundwater</u> in <u>sandstone</u> or other porous <u>rock</u>.

Lime A substance containing mainly calcium oxide (CaO) or calcium hydroxide ($Ca(OH)_2$).

Limestone A sedimentary rock chiefly composed of the <u>mineral</u> <u>calcite</u>. It is often <u>chemically precipitated</u>, but it may be <u>clastic</u> instead.

Lippo The author's term for a chimerical lion-hippo hybrid. Apparently its habitat is severely restricted; so far it has only been spotted inhabiting the <u>façade</u> of Milwaukee's St. Joseph's Convent Chapel.

Lithified Referring to <u>sediments</u> that have been turned into solid <u>rock</u> through compaction or natural cementation.

Lithology The description of a <u>rock</u>'s or rock unit's characteristics and identification traits.

Loess Windblown <u>silt</u> that originally had been deposited by glacial meltwater.

Loggia A room or porch with an open, colonnaded front.

Lower When applied to geologic time units, it denotes the oldest or earliest portion of that unit. For example, Lower Jurassic refers to the earliest portion of the <u>Jurassic period</u>.

Low-Grade Metamorphic Rock A <u>metamorphic</u> <u>rock</u> that has been subjected to a relatively gentle increase in pressure, and to temperatures between 200 and 320 degrees Celsius (392 to 608 degrees Fahrenheit).

Macroscopic Visible to the naked eye; not requiring magnification.

Mafic Referring to dark-colored <u>igneous</u> <u>rocks</u> that are rich in iron and magnesium. Such rocks include <u>basalt</u> and <u>gabbro</u>.

Magma Molten <u>rock</u> below the Earth's surface.

Magnetite An <u>oxide mineral</u> with the chemical formula Fe_3O_4. It is one of the main <u>ores</u> of iron and is well known for its magnetic properties.

Mansard Roof A roof that features two different slopes, with the lower being steeper than the upper.

Mantle The zone of the Earth's interior, about 1,800 mi (2,900 km) deep, that lies directly under the <u>crust</u>.

Mantle Plume A large column of superheated magma that rises through the mantle and, if it reaches the surface, creates a hot spot.

Marble (Architecture) An imprecise and variably defined term that generally refers to any carbonate and some noncarbonate rock types that are relatively soft and can take a high polish. Examples include true marble, limestone, serpentinite, breccia, and ophicalcite.

Marble (Geology) The metamorphic equivalent of such carbonate rocks as limestone and dolostone. It has been subjected to high temperatures, and is widely used in sculpture and cladding because it takes a high polish.

Marshfield Terrane The microcontinent that collided with the Superior Craton during the final phase of the Penokean Orogeny.

Masonry The use of dimension stone, brick, concrete blocks, or other hard materials for wall and building construction.

Mazatzal Orogeny An orogeny that occurred at about 1.65 to 1.60 Ga when an island arc collided with the underside of Laurentia.

Mazatzal Terrane A terrane composed of rocks 1.7 to 1.6 Ga old that accreted to Laurentia in a region in the US that now stretches from the Upper Midwest to the Southwest.

MCR The abbreviation for Midcontinent Rift.

Medium-Grained Referring to igneous or metamorphic stone types whose feldspar crystals are between 0.5 and 1.0 cm in length.

Meguma A terrane that apparently originated in Gondwana but subsequently parted from it and still later collided with eastern Canada. It now forms the southern portion of Nova Scotia.

Mesoarchean Era The span of geologic time from 3.2 to 2.8 Ga.

Mesoproterozoic Era The span of geologic time from 1.6 to 1.0 Ga.

Mesozoic Era The span of geologic time from 252 to 66 Ma.

Metamorphic Referring to any rock that has been subjected to increased temperature, increased pressure, or both, and therefore has been transformed into a type different from its original form.

Metaquartzite A quartzite of metamorphic origin.

Meteorite An object of interplanetary origin that reaches the surface of the Earth without completely burning up in the atmosphere.

Mica A term for the family of rock-forming silicate minerals that exhibit flat, sheetlike crystalline structure.

Microcline An alkali feldspar mineral of the composition $KAlSi_3O_8$.

Microcontinent A detached section of continental crust that is smaller than a full continent.

Microcrystalline Referring to crystals that are too small to be seen with the naked eye.

Microscopic Not visible to the naked eye; requiring magnification.

Midcontinent Rift The large, arc-shaped rift situated from Oklahoma through the Lake Superior region to Alabama that formed in the Mesoproterozoic era, from approximately 1.1 to 1.0 Ga.

Migmatite A rock that is considered a hybrid type because it is composed of both igneous and metamorphic components. One example would be a gneiss that contains veins or other inclusions of granite.

Mineral A naturally occurring, inorganic substance that has a definite chemical composition. Many minerals form crystals.

Mineralogy The study of rock-forming minerals.

Miocene Epoch The span of geologic time from 23 to 5 Ma.

Mississippian Subperiod The subdivision of the Carboniferous period from 359 to 323 Ma.

Mollusk An invertebrate animal of the phylum Mollusca. Mollusks include snails and cephalopods.

Monzogranite A granite that has between 35 and 65 percent of its total feldspar content in the form of plagioclase.

Monzonite A granitoid rock containing approximately equal amounts of plagioclase and alkali feldspars, and 5 percent or less of quartz.

Moraine *See* End Moraine.

Mortar A substance used by masons to bind dimension stones or bricks together at their joints. While the composition of mortars has varied over the centuries, modern mortar characteristically contains Portland Cement, sand, extra lime, and water.

Moulin Kame A kame formed from the accumulation of outwash in a glacier's drainage hole or channel.

Muscovite A light-colored, sheet-forming silicate mineral of the mica family. One of its listed chemical formulas is $KAl_2(Si_3Al)O_{10}(OH,F)_2$; other variations are cited.

Nappe A mass of rock that, during crustal compression, is thrust over other rock units.

Narthex That portion of a church interior just inside the entranceway or vestibule.

Natural Cement Cement that can be directly produced from argillaceous carbonate rock without the need for additional ingredients.

Nautiloid A member of an ancient line of cephalopods that was most prevalent in the Paleozoic era. One surviving species of this group is the chambered nautilus.

Nave The main interior section of a church between the apse and the narthex.

Negative Feedback Loop A process in which an initial condition or effect triggers its own mitigation or diminishment.

Nektonic Referring to marine organisms that actively swim rather than drift or dwell on the seafloor.

Neoarchean Era The span of geologic time from 2.8 to 2.5 Ga.

Neoclassical Referring to the architectural style employed from the Renaissance to modern times that is based on the architectural orders and elements of ancient Greece and Rome.

Neoproterozoic Era The span of geologic time from 1.0 Ga to 541 Ma.

Nodular Containing nodules or having a lumpy or blotchy appearance.

Nonfoliated A term applied to metamorphic rocks that do not have minerals in a parallel alignment. As a result, they do not have a banded, wavy, or thinly layered appearance.

Norman Brick Brick that is longer but flatter than the standard American size. Its dimensions are often cited as 3 5/8 x 2 1/4 x 11 5/8 in. It is virtually indistinguishable from Roman Brick.

Oceanic Crust The part of the Earth's crust that floors ocean basins. This crust is largely composed of mafic rocks.

Oligocene Epoch The span of geologic time from 34 to 23 Ma.

Olivine A silicate mineral notable for its green color. Its chemical formula is $(Mg^{2+}, Fe^{2+})_2SiO$.

Onyx A form of the mineral chalcedony characterized by light and dark bands.

Oolite A term referring to a unusual type of sedimentary rock composed of tiny, rounded grains with concentric layers of calcite. It also sometimes contains a significant amount of iron-oxide minerals that make it an exploitable source of iron ore. Oolites often form in shallow, high-energy marine environments.

Ophicalcite A recrystallized metamorphic rock containing carbonate minerals and brecciated serpentinite.

Ordovician Period The span of geologic time from 485 to 444 Ma.

Ore A mineral or rock type, usually metallic, that can be mined at a profit.

Orogenesis The process of mountain building.

Orogeny A mountain-building event.

Orthoclase A common alkali-feldspar mineral: $K(AlSi_3O_8)$.

Orthoconic Referring to nautiloids that formed long and straight (rather than coiled) shells.

Orthoquartzite A quartzite of sedimentary origin that has not undergone metamorphism.

Outcrop An exposure of bedrock.

Outwash A deposit of sorted sand, pebbles, and cobbles carried and then deposited by glacial meltwater streams.

Overburden Any loose material that sits atop the bedrock: soil, unconsolidated sediments, etc. Also, any bedrock that lies over a desired vein of coal or ore.

Oxide Mineral A member of a chemical group of minerals that contain, among other substances, O_2.

Paleoarchean Era The span of geologic time from 3.6 to 3.2 Ga.

Paleocene Epoch The span of geologic time from 66 to 56 Ma.

Paleoproterozoic Era The span of geologic time from 2.5 to 1.6 Ga.

Paleosol An ancient soil that has been preserved as a stratigraphic unit or layer in the rock record.

Paleozoic Era The span of geologic time from 541 to 252 Ma.

Palmette A type of classical ornament that resembles a stylized palm frond.

Pan-African Orogeny The Upper Neoproterozoic mountain-building event that resulted from the collision of two large plates that formed Gondwana.

Pangaea (sometimes rendered Pangea) The supercontinent that existed from approximately 300 to 200 Ma.

Paonazzo An alternative spelling of Pavonazzo.

Parent Rock The original rock type from which a metamorphic rock forms.

Pavonazzo A highly brecciated, purple-veined variety of Carrara Marble.

Pebble A rock particle 2 to 64 mm (.08 to 2.5 in) in diameter.

Pediment The usually broad gable or squat triangular element, very common in classical architecture, set above the entablature.

Pembine-Wausau Terrane The island arc that collided with the Superior Craton during the initial phase of the Penokean Orogeny.

Pennsylvanian Subperiod The subdivision of the Carboniferous period from 323 to 299 Ma.

Penokean Orogeny A Paleoproterozoic mountain-building event affecting what is now Wisconsin. It was triggered by collisions of the Superior Craton with the Pembine-Wausau Terrane and then the Marshfield Terrane.

Peridotite A dense, dark-toned, ultramafic rock thought to originate in the Earth's mantle.

Period On the geologic time scale, the largest subdivision of an era.

Permian Period The span of geologic time from 299 to 252 Ma.

Perthite A silicate mineral composed of the intergrowth of an alkali feldspar and a plagioclase feldspar.

Petrology The study of a rock's chemical properties, mineralogical content, and environment of formation.

Phaneritic Composed a macroscopic crystals.

Phenocryst A large crystal in an igneous rock that is set in a groundmass.

Pier A column or other vertical element that bears a structural load.

Pilaster A flat-faced column attached to the wall behind it. It is an ornamental rather than a load-bearing element.

Pile A pole or column driven into the substrate to provide support and stability for a structure above it.

Plagioclase Feldspars One of the two main groups of feldspar minerals. Its members, which form a continuous series, are albite, oligoclase, andesine, labradorite, bytownite, and anorthite. Plagioclase feldspars are sometimes called "soda-lime feldspars" instead.

Plaster A substance composed of lime, gypsum, or cement mixed with water that is applied to walls and other surfaces. Once dried, it forms a decorative or protective coating.

Plate One of about fifteen rigid sections that make up the Earth's lithosphere. A plate can contain oceanic crust only, or both oceanic and continental crust.

Plate Tectonics The theory first developed in the middle of the twentieth century that explains various geological phenomena in the context of moving plates.

Pleistocene Epoch The span of geologic time from 2.6 Ma to 12 ka.

Plinth The base of a pedestal, column, or statue. Also, the basal part of a building's exterior.

Pliocene Epoch The span of geologic time from 5 to 2.6 Ma.

Plug-and-Feather Method A traditional method of splitting stone by inserting two shims (the "feathers") into a hole drilled in the rock, and then hammering a wedge (the "plug") between them.

Pluton A mass of intrusive igneous rock that was originally emplaced underground.

Polished Finish A finish created by treating a stone surface with abrasives in a way that produces a highly reflective surface that emphasizes the stone's natural colors.

Poorly Sorted Referring to a sediment in which the particles vary considerably in size. In engineering parlance, this term can mean the opposite (the particles are of one uniform size). But in this book the primary, geological definition is used. *Compare* Well-Sorted.

Porphyritic Referring to an igneous rock that has phenocrysts set in a groundmass.

Porphyry An igneous rock containing large phenocrysts embedded in a groundmass.

Portland Cement The most frequently used modern cement in construction practices. It is produced by firing limestone and clay, grinding the product, called clinker, into a powder, and adding to it a small amount of gypsum.

Positive Feedback Loop A process in which an initial condition or effect triggers its own further increase or intensification.

Potassium Feldspars The subgroup of the alkali feldspars that comprises orthoclase, microcline, and sanidine.

Pozzolana Volcanic ash, originally mined by the ancient Romans in the Pozzuoli, Italy, area, that is used in the formulation of some types of cement.

Prairie School Style A style of architecture employed by Frank Lloyd Wright, George Elmslie, and other prominent Midwestern architects in the late nineteenth and early twentieth centuries. The style emphasizes a horizontality that reflects the idealized American prairie landscape and often features low, overhanging roofs and Roman Brick.

Proterozoic Eon The span of geologic time from 2.5 Ga to 541 Ma.

Protolith The original, "parent" form of a rock before it becomes metamorphosed.

Pyroxene A term referring to a group of chemically related and often dark-colored silicate minerals.

Quarry An open-air facility or enclosed-shaft mine where the local bedrock is extracted.

Quarry-Faced Stone Synonymous with rock-faced stone.

Quartz An extremely prevalent mineral composed of silicon and oxygen. Classified as either a silicate or oxide mineral, its chemical formula is SiO_2.

Quartzite Either a metamorphic equivalent of sandstone (metaquartzite), or an unmetamorphosed sandstone with grains cemented with silica (orthoquartzite).

Quartz Monzonite A granitoid rock containing approximately equal amounts of plagioclase and alkali feldspars, and between 5 and 20 percent quartz.

Quaternary Period The span of geologic time from 2.6 Ma to the present.

Queen Anne Style A style of largely domestic architecture noted for its eclectic blend of other earlier styles and its wide variety of building materials.

Quoin The masonry that forms the corner of a building.

Radiometric Age, Radiometric Date The age of a rock or other geologic specimen determined by measuring the amount of decay of an unstable isotope it contains.

Raft Foundation A raftlike concrete pad set in the shallow substrate, on which a building's load-bearing piers or walls are constructed. This construction technique generally predates the time when larger buildings were more effectively anchored with piles or caissons. However, this method has now been revived for particularly waterlogged sites.

Rebar A contraction of "reinforcing bar," as in the steel bars set in concrete to strengthen it.

Reef A term with various definitions, but in paleontology and geology it is defined as a marine structure largely made by carbonate-mineral-secreting organisms.

Reefal Referring to rock formed in a reef environment.

Reef-controlled Hill A hill or mounded landform composed of hard, erosion-resistant reefal rock.

Residual Clay Clay formed in place by the weathering of bedrock or the chemical alteration of soils.

Richardsonian Romanesque An architectural style in favor in the late nineteenth century that features such medieval Romanesque elements as semicircular arches, rock-faced stone surfaces, conical towers, and a massive, fortresslike appearance. Named for its formulator, the American architect Henry Hobson Richardson (1838–1886), it also often employs somber, dark-toned brownstone exteriors.

Ripple Marks Rippled patterns, originally made in sediments by flowing water, surf action, or the wind, that have become lithified and preserved in sandstone and other sedimentary rocks.

Riprap Large rock fragments used to stabilize a shoreline or to help stabilize a slope or hillside.

Riser The vertical section of a step or stair.

Rock A consolidated assemblage of one or more types of minerals.

Rock Caisson A deep caisson that extends downward all the way to the bedrock.

Rock-Faced Stone A form of dimension stone in which the joints of the stone are chiseled away, but the central portion is left with a jagged, rough, projecting surface.

Rockhound A superior sort of person; one who recognizes that the collecting, scientific study, and aesthetic appreciation of rock types is a primal, necessary, and transcendent human activity. Not to be confused with those credulous souls who keep "pet rocks" in little cardboard houses or stake their well-being on "mystical healing crystals."

Rodinia The supercontinent that included most modern continents and that existed from approximately 1.1 Ga to 750 Ma.

Roman Brick Brick that resembles the long, shallow brick type used by the ancient Romans. Its dimensions are often cited as 3 5/8 x 1 5/8 x 11 5/8 in, or thereabouts. Especially favored by such Prairie Style architects as Sullivan, Wright, and Elmslie, it tends to emphasize a building's horizontal lines.

Romanesque A medieval architectural style that characteristically features such elements as semicircular arches, flamboyant columns, massive towers, and barrel vaults.

Romanesque Moderne The author's flippant and unnecessary term for a building that has a Richardsonian Romanesque plinth, upon which was later erected an Art Moderne superstructure.

Rubbed Finish A finish that is produced by rubbing the stone surface with sand or some other hard substance.

Rubble Architectural stone that has not been "squared off" (cut with sides meeting at right angles). Rubble walls are usually uncoursed.

Rudist An unusual type of reef-building bivalve that existed from the Upper Jurassic period through the Cretaceous period.

Rusticated Stone Stonework in which the joints are recessed or chamfered, so that the rest of the stone surface projects outward.

Rustic Monument A type of cemetery monument sculpted from Salem Limestone that was especially popular in the 1890s and 1900s. Most rustic monuments depict dead or broken tree trunks festooned with other symbolic elements.

Saline Containing salt—often specifically the mineral halite.

Sand A rock or mineral particle .06 to 2 mm (.0024 to .08 in) in diameter.

Sandstone A clastic rock composed of sand grains that have been cemented together with silica, calcite, or another mineral.

Sauk Sequence The sequence dating from the late Neoproterozoic era to the early Ordovician period.

Scagliola An ornamental, polished plaster surface composed of gypsum, isinglass, alum, and coloring agents. It is used by architects as a relatively inexpensive stand-in for marble.

Scaling Weathered pieces that flake or peel off a stone surface.

Scalpellino (pl. Scalpellini) The Italian term for a stonecutter or stonemason.

Seam-Faced Stone A form of dimension stone in which the side of the stone that was already naturally weathered before it was quarried is used as the external face on a building. This gives the wall or façade a nicely aged and venerable appearance.

Seasoning A quarryman's term for the process of drying out porous rock such as sandstone, especially if it is extracted from below the water table. This is usually accomplished by just letting the stone stand for a prolonged period in a dry place.

Second Empire Style An architectural style current in the nineteenth century that features, among other characteristic elements, mansard roofs.

Sediment Unconsolidated material made of rock or mineral fragments.

Sedimentary Referring to any layered rock type formed either by clastic particles or by chemical precipitation. Examples include breccia, chert, conglomerate, dolostone, limestone, sandstone, and shale.

Sequence A stratigraphic term referring to a large assemblage of marine sedimentary rocks bounded above and below by unconformities. Such a deposit represents a span of geologic time when sea level rose and saltwater seas covered a large portion of continental interiors.

Sericite A term for particles of light-colored micas that produce a silky appearance.

Serpentine A term for a group of silicate minerals that are often pale to dark green. They are usually produced in metamorphism.

Serpentinite An exotic, serpentine-containing metamorphic rock whose parent rock is dunite. Usually deep green with snaking white veins, this rock is thought to originate near the Earth's crust-mantle boundary or in subduction zones.

Sessile Referring to organisms that live attached to the seafloor or other surface.

Sgraffito The artistic method of decorating walls by applying plaster layers of different tints and subsequently scoring or scratching out sections of them to reveal underlying colors that produce the desired pattern.

Shale A clastic sedimentary rock composed of clay-sized particles.

Sheet Pile A unit of steel or other impervious material designed to be driven into the substrate and connected with others to form a waterproof barrier.

Silica Synonymous with quartz.

Silicate Mineral A member of a chemical group of minerals that contains, among other substances, SiO$_4$. The one exception, quartz, is SiO$_2$.

Siliceous Containing silica.

Sill (Architecture) The shelflike, horizontal unit at the base of a window or door.

Sill (Geology) A horizontal or slanting body of intrusive rock that forms when upward-moving magma pushes its way between two strata.

Silt A mineral particle 0.003 to 0.06 mm (0.0001 to 0.003 in) in diameter.

Siltstone A clastic, sedimentary rock composed of silt particles.

Silurian Period The span of geologic time from 444 to 419 Ma.

Slab Rollback The process by which the steepening of a subducting oceanic plate causes stretching and thinning of the overlying continental crust.

Slant Marker A low gravestone with its back higher than its front, and an inscription surface angled back from the vertical so that it is more easily read by a person standing in front of it. It is sometimes called a "pillow marker" or "bevel marker" instead.

Slate The foliated, low-grade metamorphic equivalent of shale.

Soil The living skin of the Earth's land surfaces, composed of three crucial components: mineral matter, humus, and living organisms.

Spall To flake off, fall off, or splinter from a mass of rock.

Spandrel A horizontal panel that separates the windows of one story from those of the story above or below it.

Sparite A coarsely crystalline form of the carbonate mineral calcite.

Speleothem A catchall term for any cave deposits (including stalactites and stalagmites) formed by dripping or flowing water.

Spread Foundations A network of separate, detached mats or flat bases set in the shallow substrate to which a building's vertical supports are attached.

Stainless Steel A corrosion-free form of steel that is composed of iron and chromium rather than iron and carbon. Stainless steel may also contain nickel and molybdenum.

Statuario A term for the highly prized variety of Carrara Marble that has little or no veining and is pure white, or almost so. It is also more translucent than other varieties and has a texture that some sources describe as "waxy."

Steel A metal alloy most commonly composed of iron and carbon. However, stainless steel is usually a mixture of iron and chromium instead.

Stone (Architecture) Any rock type used for decorative purposes that cannot take a high polish.

Stone (Geology) Essentially synonymous with rock; it can refer to any igneous, sedimentary, or metamorphic rock type regardless of its properties, origin, or architectural use.

Stratigraphy The subdiscipline of geology that is the study, classification, and dating of rock strata.

Stratum (Plural = Strata) A layer or bed of sedimentary rock.

Stretcher A brick or block of dimension stone in ashlar masonry that is set with its long face exposed.

Stringcourse A horizontal course of masonry on a wall that is usually quite distinct from what lies below and above it.

Stromatoporoid A type of sponge that was a notable reef-builder in the Lower Paleozoic era.

Stylolite A wavy or jagged linear pattern found in limestone and other rock types. It contrasts in color with the surrounding rock and is composed of mineral or organic matter. Stylolites are often considered an ornamental asset in decorative stone.

Subduction The process in which one plate sinks beneath another.

Substrate The soil, sediment, or rock that lies directly under the Earth's surface.

Sugared Denoting the rough, granular, uneven or pitted surface of badly <u>weathered</u> <u>marble</u>.

Sulfide Mineral A member of a chemical group of <u>minerals</u> that contain, among other substances, -S (sulfur).

Supercontinent A giant landmass formed by the collision of two or more continents.

Superior Craton One of the <u>Archean crustal</u> sections that formed the ancient core of <u>Laurentia</u>. It is located in central Canada and in the uppermost part of the North-Central US.

Swell-and-Swale Terrain <u>Terrain</u> characteristic of the upper surface of an <u>end moraine</u>. It consists of hummocks, depressions, and relatively gentle slopes caused by differential settling of the moraine's <u>till</u>.

Syenite A <u>felsic</u> and <u>intrusive</u> <u>igneous</u> <u>rock</u> similar in appearance to <u>granite</u> but with significantly less <u>quartz</u> content.

Taconic (or Taconian) Orogeny The <u>orogeny</u> that occurred in the <u>Ordovician</u> and early <u>Silurian periods</u> when a volcanic <u>island arc</u> collided with <u>Laurentia</u>.

Tectonic Referring to the movement, deformation, or structural changes of the Earth's <u>crust</u>.

Tephra A general term for any airborne particles or fragments ejected from a volcano, regardless of their size.

Terra-Cotta Molded and fired <u>clay</u> used as architectural ornament (in glazed form) and for roofing tiles (either unglazed or glazed).

Terrain The surface expression of a landscape; the "lie of the land."

Terrane A three-dimensional section of the Earth's <u>crust</u> that is bounded by <u>faults</u> on all sides and that is composed of <u>rock</u> units that share a common origin.

Tertiary Period The span of geologic time from 66 to 2.6 Ma.

Tethys Ocean (Tethys Sea, Tethys Seaway) A major body of water, floored by <u>oceanic crust</u>, that formed in the <u>Mesozoic era</u>. Its much smaller modern remnants are the Mediterranean, Black, Caspian, and Aral Seas.

Thermal Finish A <u>finish</u> created by treating a <u>stone</u> surface with a high-temperature flame from a blowpipe. This produces a rough, nonskid texture.

Till The form of <u>drift</u> that is deposited either under or directly in front of a <u>glacier</u>. It is composed of all <u>sediment</u> sizes, from <u>clay</u> and <u>silt</u> to <u>gravel</u> and <u>boulders</u>. For that reason it is considered unsorted <u>drift</u>.

Tonalite A <u>felsic</u> or <u>intermediate</u>, <u>intrusive igneous rock</u> that has its <u>feldspar</u> content mostly in the form of <u>plagioclase feldspar</u>, with 10 percent or less <u>alkali feldspar</u>; also, its <u>quartz</u> content is at least 20 percent. It is thought that tonalites form from the melting of <u>mafic oceanic crust</u> when it is subducted at a <u>convergent plate boundary</u>.

Tooled Finish A <u>finish</u> created by treating a <u>stone</u> surface with a large, single-pointed hammering tool. The result is similar to a <u>bush-hammered finish</u>, but it can be selectively applied to a small area.

Topographic Referring to <u>topography</u>.

Topography The disposition of <u>landforms</u> and other natural features of a particular area; the "lie of the land."

Tosa The local nickname for Wauwatosa, Wisconsin.

Trace Fossil A fossil that provides clues about its organism's behavior, locomotion, or feeding habits rather than about its actual appearance.

Trachyandesite An <u>extrusive</u>, <u>intermediate</u> <u>igneous</u> <u>rock</u>.

Travertine A <u>limestone</u> formed by the deposition of <u>calcite</u> in hot springs or in caves where ambient-temperature water drips to create <u>speleothems</u>.

Tread The horizontal section of a step or stair.

Triassic Period The span of geologic time from 252 to 251 Ma.

Trilobite A segmented marine <u>arthropod</u>, now extinct, that formed an important part of the <u>fossil</u> record from the <u>Cambrian</u> through the <u>Permian periods</u>.

Trim An architectural term for <u>stone</u>, <u>brick</u>, or <u>terra-cotta</u> used as a highlighting contrast to the main material of a building's exterior. Trim often takes the form of thin horizontal <u>courses</u>, window <u>sills</u>, or capstones on church buttresses.

Tudorranean The author's own, wholly unauthorized term for building designs that shamelessly mix Tudor and Mediterranean elements, such as half-timbered gables and terra-cotta-tile roofs.

Turbidite Deposit A deposit of <u>sediments</u> laid down underwater by a <u>turbidity current</u>.

Turbidity Current A fast-moving current of dense, <u>sediment</u>-laden water. It can be caused by an earthquake or underwater avalanche.

Ultramafic Referring to <u>igneous rocks</u> that are very rich in iron and magnesium. They are rarely found in the Earth's <u>crust</u> but thought to be the main constituent of much of the Earth's <u>mantle</u>. They include <u>peridotite</u> and its variant <u>dunite</u>.

Unconformity A gap in the stratigraphic record caused by an episode of <u>erosion</u> or no new formation of <u>rock</u>.

Uncoursed A term that refers to <u>masonry</u> that is set in a random pattern, and not in <u>courses</u>.

Underclay The layer of <u>clay</u> found directly beneath a <u>stratum</u> of Pennsylvanian (<u>Upper Carboniferous</u>) coal. Underclays originally formed as the <u>soils</u> in which the coal-producing vegetation was rooted.

Ungulate A hoofed mammal.

Upper When applied to geologic time units, it denotes the youngest or latest portion of that unit.

Valve As used in biology, the type of shell found in such animals as <u>bivalves</u> and <u>brachiopods</u>.

Variscan Orogeny The European mountain-building event that occurred in the late <u>Carboniferous</u> and <u>Permian</u> as part of the assembly of <u>Pangaea</u>.

Vermiculated Referring to stonework or ironwork that has been carved into intricate, curving and grooved patterns resembling worm trails.

Vestibule The relatively small room just inside an entrance that leads to a building's main interior spaces.

Victorian Gothic *See* Gothic Revival Style.

Vitreous Glassy, or made of glass.

Volcanic Arc A line of volcanic mountains formed by <u>magma</u> rising from a <u>subducting plate</u>.

Volute A decorative element of <u>capitals</u> of the <u>Ionic order</u> and the <u>Composite order</u>. It resembles the rolled-up end of a scroll.

Voussoir A wedge-shaped <u>stone</u> that serves as a section of an arch.

Vug A hole or cavity in a <u>rock</u>.

Wadi A desert riverbed, dry except for brief periods of rain.

Water Table (Architecture) A <u>course</u> of <u>masonry</u>, either at the base of a building's exterior or above it, that projects outward somewhat from the rest of the wall. A water table helps direct rainwater away from the building's <u>foundation</u>, but is sometimes used by masons and architects as a purely decorative feature.

Water Table (Geology) The plane or surface that is the boundary between the zone of aeration and the zone of saturation.

Weathering The process by which <u>rock</u>, <u>soil</u>, and other materials exposed to the elements undergo chemical or physical change due to such factors as the weather, pollution, salt compounds, or the action of organisms. While weathering sometimes produces attractive and desirable coloration in building <u>stone</u>, it can in other situations be extremely destructive.

Weathering Steel A type of <u>steel</u> that has been formulated to quickly oxidize on its surface. This outer layer serves as a sealant that retards further oxidation.

Well-Sorted Referring to a <u>sediment</u> in which the particles are of one uniform size. In engineering parlance, this term can mean the opposite (a wide range of particle sizes is present). But in this book the primary, geological definition is used. *Compare* **Poorly Sorted.**

Wisconsin Glaciation The final ice sheet advance of the <u>Pleistocene epoch</u>. It lasted from 75 ka to 12 ka.

Xenolith A piece of <u>rock</u> embedded in a mass of different, <u>igneous</u> rock. Xenoliths often form when fragments of preexisting rock fall into a body of intruding <u>magma</u>.

Yavapai Orogeny An <u>orogeny</u> that occurred at about 1.7 Ga when an island arc collided with the underside of <u>Laurentia</u>.

Yavapai Terrane A <u>terrane</u> composed of <u>rocks</u> 1.8 to 1.7 Ga old that accreted to <u>Laurentia</u> in a region in the US that stretches from the Upper Midwest to the Southwest.

Zircon An extremely nonreactive <u>silicate mineral</u> with the chemical formula $ZrSiO_4$. Zircon is very resistant to melting and <u>weathering</u> and survives geologic processes destructive to most other <u>minerals</u>.

Zopfstil The <u>neoclassical</u> architectural style favored in Germany from the late eighteenth century, and found in selected Milwaukee designs of the nineteenth century.

Zuñi Sequence The <u>sequence</u> dating from the late <u>Jurassic period</u> to the <u>Paleocene epoch</u>.

Selected Bibliography

GENERAL GEOLOGY

Braschayko, Suzanne M. *The Waukesha Fault and Its Relationship to the Michigan Basin: A Literature Compilation.* Open-File Report 2005–05. Madison: Wisconsin Geological and Natural History Survey, 2005.

Brown, B. A. *Bedrock Geology of Wisconsin, West-Central Sheet.* Regional Map Series, Survey Map 88–7. Madison: Wisconsin State Geological and Natural History Survey, 1988.

Coorough Burke, Patricia, and Peter M. Sheehan. "Museums at the Intersection of Science and Citizen: An Example from a Silurian Reef." In *Museums at the Forefront of the History and Philosophy of Geology: History Made, History in the Making,* edited by G. D. Rosenberg and R. M. Clary, 263–271. Geological Society of America Special Paper 535.

Dorr, John A., Jr., and Donald F. Eschman. *Geology of Michigan.* Ann Arbor: University of Michigan Press, 1970.

Dott, Robert H., Jr., and John W. Attig. *Roadside Geology of Wisconsin.* Missoula, MT: Mountain Press, 2004.

Gass, Kenneth C., Joanne Kluessendorf, Donald G. Mikulic, and Carlton E. Brett. *Fossils of the Milwaukee Formation: A Diverse Middle Devonian Biota from Wisconsin, USA.* Manchester, UK: Siri Scientific, 2019.

Geological Society of America. "GSA Geologic Time Scale v. 5.0." Accessed June 29, 2020. https://www.geosociety.org/documents/gsa/timescale/timescl.pdf.

Gilman, Richard A., Carleton A. Chapman, Thomas V. Lowell, and Harold W. Borns Jr. *The Geology of Mount Desert Island.* Augusta: Maine Geological Survey, 1985.

Hatch, Norman L., Jr., ed. *The Bedrock Geology of Massachusetts.* Professional Paper 1366-E-J. Washington, DC: United States Geological Survey, 1991.

Heiken, Grant, Renato Funiciello, and Donatella De Rita. *The Seven Hills of Rome: A Geological Tour of the Eternal City.* Princeton, NJ: Princeton University Press, 2005.

Higgins, Michael Denis, and Reynold Higgins. *A Geological Companion to Greece and the Aegean.* Ithaca, NY: Cornell University Press, 1996.

Howell, Paul D., and Ben A. van der Pluijm. "Early History of the Michigan Basin: Subsidence and Appalachian Tectonics." *Geology* 18 (December 1990): 1195–1198.

Kean, William K. "The Neda Iron Ore of Southeastern Wisconsin." University of Wisconsin–Milwaukee Field Station Bulletin, Fall 1986. Accessed March 11, 2023. https://dc.uwm.edu/cgi/viewcontent.cgi?article=1114&context=fieldstation_bulletins.

Kluessendorf, Joanne, and Donald D. Mikulic. "National Historic Landmark Nomination: Schoonmaker Reef." Accessed April 3, 2022. https://npgallery.nps.gov/NRHP/GetAsset/NHLS/97001266_text.

Kluessendorf, Joanne, and Donald D. Mikulic. "National Historic Landmark Nomination: Soldiers' Home Reef." Accessed April 3, 2022. https://npgallery.nps.gov/NRHP/GetAsset/NHLS/97001266_text.

LaBerge, Gene L. *Geology of the Lake Superior Region*. Phoenix: Geoscience, 1994.

Lapham, Increase Allen. *Wisconsin: Its Geography and Topography, History, Geology, and Mineralogy*. 2nd ed. Milwaukee: I. A. Hopkins, 1846.

Lasca, N. P. "Quaternary Stratigraphy of Southern Milwaukee County, Wisconsin Preliminary Results." *Geoscience Wisconsin* 7 (July 1983): 17–23.

Malone, David H., Carol A. Stein, John P. Craddock, Jonas Kley, Seth Stein, and John E. Malone. "Maximum Depositional Age of the Neoproterozoic Jacobsville Sandstone, Michigan: Implications for the Evolution of the Midcontinent Rift." *Geosphere* 12, no. 4 (2016): 1–12.

McCormick, Kelli A. *Precambrian Basement Terrane of South Dakota*. Bulletin 41. Vermilion: South Dakota Geological Survey, 2010.

Mikulic, Donald G., and Joanne Kluessendorf. *The Classic Silurian Reefs of the Chicago Area*. ISGS Guidebook 29. Champaign: Illinois State Geological Survey, 1999.

Mikulic, Donald G., and Joanne Kluessendorf. "Subsurface Stratigraphic Relationships of Upper Silurian and Devonian Rock of Milwaukee County, Wisconsin." *Geoscience Wisconsin* 12 (July 1988).

Milwaukee Public Museum. Virtual Silurian Reef (website). https://www.mpm.edu/content/collections/learn/reef/grafton-front.html.

Sampsell, Bonnie M. *The Geology of Egypt: A Traveler's Handbook*. Cairo: American University in Cairo, 2014.

Schneider, Allan F. "Wisconsinan Stratigraphy and Glacial Sequence in Southeastern Wisconsin." *Geoscience Wisconsin* 7 (July 1983): 59–85.

Schulz, Klaus J., and William F. Cannon. "The Penokean Orogeny in the Lake Superior Region." *Precambrian Research* 157 (2007): 4–25.

Scotese, Christopher R. Paleomap Project (website). http://www.scotese.com.

Stein, Seth, Carol A. Stein, Reece Elling, Jonas Kley, G. Randy Keller, Michael Wysession, Tyrone Rooney, Andrew Frederiksen, and Robert Moucha. "Insights from North America's Failed Midcontinent Rift into the Evolution of Continental Rifts and Passive Continental Margins." *Tectonophysics* 744 (2018): 403–421.

Stein, S., C. Stein, J. Kley, R. Keller, M. Merino, E. Wolin, D. Wiens, et al. "New Insights into North America's Midcontinent Rift." *Eos*, August 4, 2016. Accessed April 2, 2022. https://eos.org/features/new-insights-into-north-americas-midcontinent-rift.

Taylor, Simon, and Stephen Allard. "Geochemical Classification of the East Central Minnesota Batholith." A poster presented at the Ramaley Research Conference, April 4, 2021, Winona State University, Winona, MN. Accessed March 19, 2022. https://openriver.winona.edu/cgi/viewcontent.cgi?article=1033&context=wsurrc.

Thwaites, Fredrik Turville. *Sandstones of the Wisconsin Coast of Lake Superior*. Bulletin 25. Madison: Wisconsin State Geological and Natural History Survey, 1912.

Whitmeyer, Steven J., and Karl E. Karlstrom. "Tectonic Model for the Proterozoic Growth of North America." *Geosphere* 3, no. 4 (August 2007): 220–259.

United States Geological Survey. Mineral Resources On-Line Spatial Data (website): mrdata.usgs.gov.

ARCHITECTURAL STONE TYPES AND QUARRYING

Architectural Conservation Laboratory at the University of Pennsylvania School of Design. "The Slate Belt." Accessed August 31, 2021. https://sites.google.com/view/theslatebelt/introduction/the-slate-belt.

Austin, Muriel B., Arthur M. Hussey II, and John R. Rand. *Maine Granite Quarries and Prospects*. Maine Geological Survey Mineral Resources Index No. 2. Augusta: Maine Department of Economic Development, 1958.

Babcock-Smith House Museum. "Granite Industry, Westerly, RI." Accessed September 24, 2021. https://www.babcocksmithhouse.org/GraniteIndustry/index.htm.

Barron, A. J. Mark. "Carrara Marble." *Mercian Geologist* 19 (October 2018): 188–194. Accessed August 3, 2020. https://www.researchgate.net/publication/338480228_Carrara_Marble.

Bowles, Oliver. *The Stone Industries*. 2nd ed. New York: McGraw-Hill, 1939.

Bowles, Oliver. *The Structural and Ornamental Stones of Minnesota*. US Geological Survey Bulletin 663. Washington, DC: Government Printing Office, 1918.

Bownocker, J. A. *Building Stones of Ohio*. Fourth Series, Bulletin 18. Columbus: Geological Survey of Ohio, 1915.

Buckley, Ernest Robertson. *On the Building and Ornamental Stones of Wisconsin*. Madison: Wisconsin Geological and Natural History Survey, 1898.

Buckley, Ernest Robertson, and Henry Andre Buehler. *The Quarrying Industry of Missouri*. Jefferson City: Missouri Bureau of Geology and Mines, 1904.

Dale, T. Nelson. *The Commercial Granites of New England*. United States Geological Survey Bulletin 723. Washington, DC: US Government Printing Office, 1923.

Dale, T. Nelson, et al. *The Commercial Marbles of Western Vermont*. United States Geological Survey Bulletin 521. Washington, DC: US Government Printing Office, 1912.

Dale, T. Nelson, et al. *Slate Deposits and Slate Industry of the United States*. United States Geological Survey Bulletin 275. Washington, DC: US Government Printing Office, 1906.

Dickie, G. B. "Building Stone in Nova Scotia." Information Circular 12, 3rd ed. Halifax, NS, 1993. Accessed February 3, 2022. https://novascotia.ca/natr/meb/data/pubs/ic/ic12.pdf.

Eckert, Kathryn Bishop. *The Sandstone Architecture of the Lake Superior Region*. Detroit: Wayne State University, 2000.

Eckert, Kathryn Bishop. *The Sandstone Quarries of the Apostle Islands*. A report submitted to the Apostle Isles National Lakeshore, National Park Service, 1985. Accessed March 12, 2022. http://npshistory.com/publications/apis/sandstone-quarries.pdf.

Fenneman, N. M. *Geology and Mineral Resources of the St. Louis Quadrangle, Missouri-Illinois*. Bulletin 438. Washington, DC: US Geological Survey.

Fick, Charles, and Kenneth Jones. *Amberg Quarries Scrapbook*. Amberg, WI: Amberg Historical Society, 2020.

Fiddes, Jim. *The Granite Men: A History of the Granite Industries of Aberdeen and North East Scotland*. Stroud, UK: History Press, 1913.

Georgia Marble Company. *Yesterday, Today, and Forever: The Story of Georgia Marble*. Tate, GA: Georgia Marble Company, n.d.

Gordon, Charles Henry. *The Marbles of Tennessee*. Nashville: Tennessee State Geological Survey, 1911.

Hawes, George W., et al. *Report on the Building Stones of the United States, and Statistics of the Quarry Industry for 1880*. Washington, DC: US Census Office, 1883.

Hebrank, Arthur H. *The Geologic Story of the St. Louis Riverfront (A Walking Tour)*. Special Publication 6. Rolla: Missouri Department of Natural Resources, Division of Geology and Land Survey, 1989.

Heinrich, E. William. *Economic Geology of the Sand and Sandstone Resources of Michigan.* Lansing: Michigan Department of Environmental Quality Geological Survey Division, 2001.

Hieb, James. *Sandstone Center of the World.* South Amherst, OH: Quarrytown.net, 2007.

Howe, John Allen. *The Geology of Building Stones.* London: Routledge, 1910.

Indiana University. "Building a Nation: Indiana Limestone Photograph Collection." Accessed March 20, 2020. http://webapp1.dlib.indiana.edu/images/splash. htm?scope=images/VAC5094.

Jenkins, Joseph. *The Slate Roof Bible.* 3rd ed. Grove City, PA: Joseph C. Jenkins, 2016.

Lamar, J. E., and H. B. Willman. *Illinois Building Stones.* Report of Investigations 184. Urbana, IL: Division of the State Geological Survey, 1955.

Lent, Frank A. *Trade Names and Descriptions of Marbles, Limestones, Sandstones, Granites and Other Building Stones Quarried in the United States, Canada, and Other Countries.* New York: Stone, 1925.

Loerke, Jean Penn. *Waukesha Limestone: The Quarries, the Kilns, and the Buildings.* Waukesha, WI: Waukesha County Historical Museum, 1978.

McClymont, J. J. *A List of the World's Marbles.* Farmington, MI: Marble Institute of America, 1990.

McDonald, William H. *A Short History of Indiana Limestone.* Bedford, IN: Lawrence County Tourism Commission, 1995.

Merrill, George Perkins. *Stones for Building and Decoration.* 3rd ed. New York: John Wiley & Sons, 1908.

National Slate Association. *Slate Roofs.* 2010 ed. Poultney, VT: National Slate Association, 2010.

North Country Slate. Fact sheets listing buildings with roofing slate provided by North Country Slate. Supplied to the author by North Country Slate on February 7, 2019.

Ohio Division of Geological Survey. "Ohio's Sandstone Industry." *Ohio Geology Newsletter,* Spring 1982.

Oxford Museum of Natural History. "Corsi Collection of Decorative Stones." Accessed May 17, 2019. http://www.oum.ox.ac.uk/corsi/.

Parks, William A. *Report on the Building and Ornamental Stones of Canada.* Vol. 2, *The Maritime Provinces.* Ottawa: Government Printing Bureau, 1914.

Patton, John B., and Donald D. Carr. *The Salem Limestone in the Indiana Building-Stone District.* Occasional Paper 28. Bloomington: Indiana Geological Survey, 1982.

Pepper, James F., Wallace De Witt Jr., and David F. Demarest. *Geology of the Bedford Shale and Berea Sandstone in the Appalachian Basin.* US Geological Survey Professional Paper 259. Washington, DC: US Government Printing Office, 1954.

Perazzo, Peggy B. Stone Quarries and Beyond (website). https://quarriesandbeyond. org/index.html.

Pivko, Daniel. "Natural Stones in Earth's History." *Acta Geologica Universitatis Comnenianae* 58 (2003): 73–86.

Pullen, H. W. *Handbook of Ancient Roman Marbles.* London: John Murray, 1894.

Renwick, William George. *Marble and Marble Working.* New York: Van Nostrand, 1909.

Stauffer, Clinton R. *The High Magnesium Dolomites and Dolomitic Limestones of Minnesota.* Summary Report No. 4. Minneapolis: Minnesota Geological Survey, 1950.

Stone Contact (website). https://www.stonecontact.com/.

Stone, Ralph. *Building Stones of Pennsylvania.* Bulletin M 15. Harrisburg, PA: Topographic and Geologic Survey, 1932.

Stout, Wilber. "Sandstones and Conglomerates in Ohio." *Ohio Journal of Science* 44, no. 2 (1944): 75–88.

Thiel, George A., and Carl E. Dutton. *The Architectural, Structural, and Monumental Stones of Minnesota.* Minneapolis: University of Minnesota Press, 1935.

Troost, Bryan, Kathleen McGwin, and Thomas Freitag. *Etched in Stone: History of the Montello Granite Quarries.* 2nd ed. Montello, WI: Kathleen McGwin, 2018.

Winkler, Erhard M. *Stone in Architecture: Properties, Durability.* 3rd ed. Berlin: Springer-Verlag, 1997.

OTHER BUILDING MATERIALS

Ambrose, Pamela, Mimi Stiritz, and Joseph Heathcott. *Brick by Brick: Building St. Louis and the Nation.* St. Louis: St. Louis University, 2004.

Berry, George A. III, with Sharon S. Darling. *Common Clay: A History of American Terra Cotta Corporation, 1881–1966.* Crystal Lake, IL: TCR, 2003.

Buckley, Ernest Robertson. *The Clays and Clay Industries of Wisconsin.* Madison: Wisconsin Geological and Natural History Survey, 1901.

Campbell, James W. P. *Brick: A World History.* London: Thames & Hudson, 2003.

Courland, Robert. *Concrete Planet.* Amherst, NY: Prometheus Books, 2011.

Gayle, Margot, David W. Look, and John G. Waite. *Metals in America's Historic Buildings.* Rev. ed. Washington, DC: National Park Service, 1992.

Ladd, G. E. "The Clay, Stone, Lime and Sand Industries of St. Louis City and County." In *Bulletin 3.* Jefferson City: Geological Survey of Missouri, 1890.

Peck, Herbert. *The Book of Rookwood Pottery.* New York: Crown, 1968.

Ries, Heinrich. *The Clays of Wisconsin and Their Uses.* Madison: Wisconsin Geological and Natural History Survey, 1906.

Ries, Heinrich, W. S. Bayley, et al. *High-Grade Clays of the Eastern United States.* Bulletin 708. Washington, DC: United States Geological Survey, 1922.

Ries, Heinrich, and Henry B. Kümmel. *The Clays and Clay Industry of New Jersey.* Vol. 6 of the Final Report of the State Geologist. Trenton, NJ: MacCrellish & Quigley, 1904.

Ries, Heinrich, and Henry Leighton. *History of the Clay-Working Industry in the United States.* New York: John Wiley & Sons, 1909.

Roesler, Robert B. *City of Milwaukee Cream City Brick.* A typewritten survey with photographs, dated 1981. Archives of the Milwaukee County Historical Society, Milwaukee.

Stern, Andrew Charles. "Cream City: The Brick That Made Milwaukee Famous." Master's thesis, University of Georgia, 2015.

ARCHITECTURE AND PUBLIC ART

Architecture of Faith. "Historic Churches and Synagogues of Milwaukee." http://architectureoffaithmilwaukee.info/Introduction/.

Charles Allis Art Museum. "The Mansion." Accessed July 17, 2022. https://www.charlesallis.org/the_mansion/?room#first-level=marble-hall.

City of Milwaukee. Historic Preservation Study Reports. https://city.milwaukee.gov/ImageLibrary/Groups/cityHPC/DesignatedReports/.

City of Milwaukee. "Milwaukee Ethnic Houses Tour." Brochure published by the City of Milwaukee Department of City Development, August 1994. Accessed

March 11, 2022. https://city.milwaukee.gov/ImageLibrary/Groups/cityHPC/
   Books/EthnicHouses-OCR11.pdf.
City of Milwaukee. "Milwaukee War Memorials; Inventory Sheet." Accessed
   February 1, 2022. https://city.milwaukee.gov/ImageLibrary/Groups/ccClerk/
   HPC/PDFs/Memorial-Inventory-Sheets/Ald.04-06-LincolnMemorialStatue-
   Lakefront.pdf.
Doors Open Milwaukee (website). https://www.doorsopenmilwaukee.org/.
Garber, Randy, ed. *Built in Milwaukee: An Architectural View of the City*. Milwaukee:
   City of Milwaukee, 1983.
Grundl, Tim, Nancy Hubbard, and Bill Kean. "Virtual Tour of Downtown Milwaukee:
   The Buildings and Building Stones of Downtown Milwaukee." Accessed
   November 8, 2018. http://people.uwm.edu/urban-geology/.
Harris, Cyril M. *Dictionary of Architecture and Construction*. 4th ed. New York:
   McGraw-Hill, 2006.
Korom, Joseph. *Milwaukee Architecture: A Guide to Notable Buildings*. Madison, WI:
   Prairie Oak, 1995.
Milwaukee Houses of Worship: 1975 Survey. Archives of the Milwaukee County
   Historical Society.
MKE Downtown. "Public Art." Accessed February 3, 2022. https://www.
   milwaukeedowntown.com/experience/public-art.
O'Gorman, James F. *Living Architecture: A Biography of H. H. Richardson*. New York:
   Simon & Schuster, 1997.
"The Old Light House Bluffs." *Milwaukee Sentinel*, February 17, 1898.
Pagel, Mary Ellen, and Virginia Palmer. *Guides to Historic Milwaukee: Juneautown
   Walking Tour*. Madison: University of Wisconsin Extension Division, 1965.
Perrin, Richard W. E. *The Architecture of Wisconsin*. Madison: State Historical Society
   of Wisconsin, 1967.
Perrin, Richard W. E. *Historic Wisconsin Buildings: A Survey in Pioneer Architecture
   1835–1870*. 2nd rev. ed. Milwaukee: Milwaukee Public Museum, 1981.
Perrin, Richard W. E. *Milwaukee Landmarks*. Milwaukee Public Museum Publications
   in History No. 9. Milwaukee: Milwaukee Public Museum, 1968.
Schulze, Franz. *Building a Masterpiece: Milwaukee Art Museum*. New York: Hudson
   Hills, 2001.
Society of Architecture Historians, SAH Archipedia (website). https://sah-archipedia.org.
Tanzilo, Robert. *Hidden History of Milwaukee*. Charleston, SC: History Press, 2014.
Tanzilo, Robert. "Urban Spelunking" series of online articles. https://onmilwaukee.
   com/articles/.
United States National Park Service. National Register of Historic Places Inventory.
   https://www.nps.gov/subjects/nationalregister/index.htm.
Weisiger, Marsha, et al. *Buildings of Wisconsin*. Charlottesville: University of Virginia
   Press, 2016.
Wisconsin Historical Society. Property Records. https://www.wisconsinhistory.org/
   Records.
Young, Mary Ellen. Handwritten and typewritten architectural notes on various
   Milwaukee buildings. Mary Ellen Young Records Collection, archives of the
   Milwaukee County Historical Society, Milwaukee.
Zimmermann, H. Russell. *The Heritage Guidebook*. 2nd ed. Milwaukee: Harry W.
   Schwartz, 1989.
Zimmermann, H. Russell. *Magnificent Milwaukee: Architectural Treasures 1850–1920*.
   Milwaukee: Milwaukee Public Museum, 1987.
Zimmermann, H. Russell. *The Milwaukee Club*. Milwaukee: Milwaukee Club, 1982.

CIVIL ENGINEERING AND CONSTRUCTION

Giles Engineering Associates. "Northwestern Mutual." Accessed March 16, 2022. https://www.gilesengr.com/northwestern-mutual/.

Hatch, James N. "Foundations for Large Buildings (Concluded)." *Inland Architect and News Record* 45, no. 5 (July 1905): 50.

Hoffmann, Donald. "Pioneer Caisson Building Foundations: 1890." *Journal of the Society of Architectural Historians* 25, no. 1 (1966): 68–71.

Levy, Matthys, and Mario Salvadori. *Why Buildings Fall Down.* Rev. ed. New York: W. W. Norton, 2002.

Manierre, George. "A Description of Caisson Work to Bed Rock for Modern High Buildings." *Journal of the Illinois State Historical Society* 9, no. 3 (1916): 296–300.

Peck, Ralph B. *History of Building Foundations in Chicago.* Engineering Experiment Station Bulletin Series No. 373. *University of Illinois Bulletin* 45, no. 29 (January 2, 1948).

Schock, Robert E., and Eric J. Risberg. "120 Years of Caisson Foundations in Chicago." *International Conference on Case Histories in Geotechnical Engineering* 6 (2013).

MILWAUKEE COUNTY HISTORY

Bergland, Martha, and Paul G. Hayes. *Studying Wisconsin: The Life of Increase Lapham.* Madison: Wisconsin Historical Society, 2014.

Chadwick, Silas. *The Forest Home Cemetery, Milwaukee, Wis.* Milwaukee: Forest Home Cemetery, n.d.

Encyclopedia of Milwaukee. Accessed March 25, 2023. https://emke.uwm.edu.

Forest Home Cemetery (Milwaukee). "Self-Guided Historical Tour." Accessed October 3, 2021. https://foresthomecemetery.com/sites/default/files/documents/FHC_Self_Tour.pdf.

Forest Home Cemetery & Arboretum Map & Guide. Milwaukee: Forest Home Cemetery, n.d.

Gurda, John. *The Making of Milwaukee.* 4th ed. Milwaukee: Milwaukee County Historical Society, 2018.

Gurda, John. *Silent City: A History of Forest Home Cemetery.* Milwaukee: Forest Home Cemetery, 2000.

Haubrich, Paul A., Robert L. Giese, Anita M. Pietrykowski, Margaret M. Berres, and Thomas A. Ludka. *Milwaukee's Forest Home Cemetery.* Charleston, SC: Arcadia, 2020.

TRADE JOURNALS AND NEWSPAPERS CONSULTED

American Architect and Building News
Architectural Forum
Architectural Record
Brick
Brick & Clay Record
Brickbuilder
Brick, Tile & Terra Cotta Workers' Journal
Building Stone Magazine
Granite
Historical Messenger of the Milwaukee Historical Society
Inland Architect and News Record

Manufacturer and Builder
Milwaukee Business Journal
Milwaukee Journal
Milwaukee Journal Sentinel
Milwaukee Sentinel
Monumental News
Stone, an Illustrated Magazine
Stone World
Through the Ages
Urban Milwaukee
Western Architect
Wisconsin Architect

Index

Page numbers in *italics* refer to figures

100 East Building, *21*, 76–78
1000 North Water Street, *14*, 109–110

Abelman Stone/Quartzite. *See* Hinckley
  Sandstone
Aberdeenshire Granite, *12*, *82*, 83–84, 85,
  187, 192
Acadian Orogeny/Mountains, 53, 67, 87,
  138, 217
Adler & Sullivan, 83
Algoman Orogeny, 109
Alicante Limestone, *12*, 78
Alkali-feldspar granite. *See* granite
Alleghenian Orogeny/Mountains, 159, 174
Allis (Sarah and Charles) House, 201–207,
  204, *206*
All Saints Episcopal Cathedral, 191
aluminum, 23, 37, 48–49, 74; bauxite principal
  ore of, 48; coated, 37, 48–49
Amberg Silver Grey Granite. *See* Athelstane
  Granite
American Architect, 81, 84, 166
American Bronze Company, 136
American Terra Cotta Company, 34, 84, 91–92,
  92, 140, 150, 208, *208*
Amherst Stone. *See* Berea Sandstone
Amoco Building. *See* Aon Center
amphibolite, 60
anorthosite, 23, 126. *See also* Isabella
  Anorthosite
Aon Center (Chicago, IL), 108
Archean eon, 11, *12*, 13, 15, 36, 109, 116, 128
Architectural Iron Works, 74
Architectural Record, 94
Arnold F. Meyer & Company, 158, 217
Art Deco style, *37*, 57, *59*, 90, 123, 124, 126,
  130–131, 132, 172, 200, 208, *208*
Art Moderne style, 57, 90–91, 95, 123, 124–125,
  132, 144, 193, 218
Athelstane Granite, *12*, *24*, *65*, 66–67, 114, 117,
  147, 200, 209
Atlantic Coastal Plain Province, 96

Atlantic Terra Cotta Company, *12*, 34, 96, *97*
Auditorium Building (Chicago, IL), 83
Avalonia, 53, 67, 87, 90, 138, 173, 192, 214

Babylonians, ancient. *See* Mesopotamians,
  ancient
Backes & Uthes, 150
Ballentine House (Waukegan, IL), 93
Baltimore, MD: as brickmaking center, 31, 67;
  great fire of 1904, 34
Banded Iron Formation, 74, 128–129, 200
Bank of Milwaukee. *See* State Bank of
  Wisconsin–Bank of Milwaukee
Baraboo Quartzite: 189
Barre Granodiorite, *12*, 24, 137–138, *138*, 173,
  173, 174, 175, 178
Baumann, Frederick, 41
bauxite, 48
Bayfield Group sandstones, 69
Beattie Limestone Formation, 89
Beaux Arts style, *52*, 53, 61, 63, 65, 89, 106, 123,
  125, 132, 139, 141, 144, 196–198, *197*
Bedford Stone/Limestone. *See* Salem
  Limestone
Bedford trees, 168–170, 171–172, 179–181, *180*
Beman, Solon S., 85, 196
Berea Sandstone, *12*, 82–83, *82*, 85, 98, 101, 111,
  132, 136, 166, 187, 215
biocalcarenite, 25, 63, 105, 132, 168, 196
"black granite," 101, 109, 126
Blatz Brewery Complex, 110–111, 112
Blatz Brewing Office Building, 111
Blatz Mausoleum, 172–173, *173*
Blue Westerly Granite. *See* Narragansett Pier
  Granite.
Blumenfeld, Helaine, 129
Boerboom, Terry, 190
book-matching, 142–143, *142*
Borgwardt Monument, 172
Boston, MA: great fire of 1872, 34
Boston Valley terra-cotta, 102
Botticino Limestone, *12*, 56–57